T0141356

About Island Press

Island Press is the only nonprofit organization in the United States whose principal purpose is the publication of books on environmental issues and natural resource management. We provide solutions-oriented information to professionals, public officials, business and community leaders, and concerned citizens who are shaping responses to environmental problems.

In 2004, Island Press celebrates its twentieth anniversary as the leading provider of timely and practical books that take a multidisciplinary approach to critical environmental concerns. Our growing list of titles reflects our commitment to bringing the best of an expanding body of literature to the environmental community throughout North America and the world.

Support for Island Press is provided by the Agua Fund, Brainerd Foundation, Geraldine R. Dodge Foundation, Doris Duke Charitable Foundation, Educational Foundation of America, The Ford Foundation, The George Gund Foundation, The William and Flora Hewlett Foundation, Henry Luce Foundation, The John D. and Catherine T. MacArthur Foundation, The Andrew W. Mellon Foundation, The Curtis and Edith Munson Foundation, National Environmental Trust, The New-Land Foundation, Oak Foundation, The Overbrook Foundation, The David and Lucile Packard Foundation, The Pew Charitable Trusts, The Rockefeller Foundation, The Winslow Foundation, and other generous donors.

The opinions expressed in this book are those of the author(s) and do not necessarily reflect the views of these foundations.

ENVIRONMENTALISM
& THE
TECHNOLOGIES OF TOMORROW

Edited by
Robert Olson &
David Rejeski

ENVIRONMENTALISM & THE TECHNOLOGIES OF TOMORROW

Shaping the Next Industrial Revolution

ISLAND PRESS
WASHINGTON • COVELO • LONDON

Chapter 2 excerpted from Lester R. Brown, *Eco-Economy: Building an Economy for the Earth* (W. W. Norton & Co., NY: 2001).

Library of Congress Cataloging-in-Publication data.

Environmentalism & the technologies of tomorrow : shaping the next industrial revolution / edited by Robert Olson, David Rejeski.
p. cm.
Includes bibliographical references and index.
ISBN 1-55963-765-X (cloth: alk. paper) — ISBN 1-55963-769-2 (pbk.: alk. paper)
1. Sustainable development. 2. Technological innovations—Environmental aspects. 3. Environmental protection. 4. Green technology. I. Olson, Robert L. (Robert Linus), 1942- II. Rejeski, David.
HC79.E5E597 2004
338.9′27—dc22

2004012440

British Cataloguing-in-Publication data available.

Printed on recycled, acid-free paper
Design by Trent Williams
Manufactured in the United States of America
10 9 8 7 6 5 4 3 2 1

We would like to dedicate this book to the many people whose work, past and present, have most deeply affected our own outlooks; and to people working within the U.S. Environmental Protection Agency today to bring greater foresight into the agency's planning and operations:

Derry Allen, Charles Ames, Walter Truett Anderson, Clement Bezold, Kenneth and Elise Boulding, Michael Brody, Gro Harlem Brundtland, Rachel Carson, Arthur C. Clarke, Henry Dreyfus, James Dator, Peter Drucker, Richard Feynman, Buckminster Fuller, William Gibson, Willis Harman, Hazel Henderson, Amory and Hunter Lovins, Jessica Mathews, Ian McHarg, Margaret Mead, Donella Meadows, Richard L. Meier, Donald N. Michael, Lewis Mumford, Dennis O'Connor, Victor Papanek, Aurelio Peccei, Jonathan Peck, John R. Platt, Renelle Rae, Mark Rothko, Debbie Rutherford, James Gustave Speth, Anita Street, Alvin and Heidi Toffler, Barbara Ward, and H.G. Wells.

Contents

Acknowledgments

A book of this kind owes more to its contributors than to its editors. We are grateful to all of them for their individual contributions and their cooperation. The book would not have been published without the support of Todd Baldwin at Island Press, who was willing to back a project likely to ruffle the feathers of people committed to traditional approaches to environmentalism and environmental protection. We would like to thank Jessica Biamonte, Alexander Genetos, and Miriam Salerno for their assistance in final editing.

This document has been reviewed by the U.S. Environmental Protection Agency (EPA) and the material changes suggested by the Agency were accommodated. This publication was developed under Cooperative Agreement No. 82955801 awarded to the Woodrow Wilson International Center of Scholars by the EPA. The EPA made comments and suggestions on the book intended to improve the scientific analysis and technical accuracy of the document. However, the views expressed in this book are those of the authors and EPA does not endorse any products or commercial services mentioned in this publication.

The editors would like to thank the EPA for their support of the production of this book. We would also like to thank the Woodrow Wilson Center Press for their review of an earlier version of the manuscript.

Introduction

Another Chance

Bob Olson and David Rejeski

Imagine, if you can, a return to the nascent days of the industrial revolution in the nineteenth century. At this critical point in history, when innovations like mass production, chemical synthesis, and steam power converged, a small group of people began to foresee possible environmental downsides associated with our emerging vision of progress. These people included visionaries like Henry David Thoreau and Ralph Waldo Emerson, social activists like Neal Ludd, and, later, conservationists and preservationists like John Muir and Gifford Pinchot. While the contributions of these forerunners of the modern environmental movement were monumental, what they lacked were the legal mechanisms and political will necessary to shape the burgeoning industrial revolution in less harmful ways. Ponder this question: What might have happened if, at that point in history, a powerful and well-organized environmental movement had arisen to help shape our emerging technological and economic infrastructure and integrate environmental concerns into the decisions of business leaders, government officials, and citizens?

That opportunity was, of course, missed. The environmental movement as we know it arose in the early 1970s and has spent much of the last thirty years dealing with the damages of a century-old revolution in industrial production. This job, ranging from the cleanup of abandoned waste sites to the modernization of our energy production infrastructure, will still take decades to finish.

Today, we are at another critical point in history, where technical changes even larger than those that produced the industrial revolution are converging. We sit at the doorsteps of multiple revolutions in production, information and communications, logistics, and the interaction of new technologies such as nano- and biotechnology. We have another chance to properly perceive these changes, integrate environmental concerns into our decision making, head off potentially serious environmental damages, and shape emerging technologies for both economic success and the health of the planet. However, the stakes are high, the tempo is fast, and the systems we are trying to influence are enormously complex in comparison to their earlier counterparts.

This collection of essays is an attempt to better understand the nature of the changes around us, their implications for the environment, and possible transition strategies to a sustainable future. We have purposely collected essays that are optimistic, because a pessimistic outcome is unacceptable and would indicate a failure of human imagination at a crucial point in our history. However, optimism must be tempered with a realistic assessment of the magnitude of the challenges we face. We not only have to learn what works environmentally in a new and emerging world but we must simultaneously unlearn assumptions and behaviors that may no longer be relevant and that may, in fact, impede our progress.

The good news is that a credible vision is emerging of how global development can continue without undermining the ecological foundations on which our economies are built. It is very different from earlier ideas about going back to simpler technologies and ways of life. It is a vision of a technologically advanced sustainable society with great institutional capabilities and a high quality of life. This desirable future is feasible because rapid technological progress occurring in many areas can allow us to sharply reduce the consumption of natural resources and the generation of pollution per unit of economic growth.

As you enter this collection, be forewarned. This is not a novel with a clear ending, but a collection of vignettes describing the social and technological transformations that are now occurring, or that will occur in the foreseeable future. The implications of these transformations for the environment and for global sustainability will depend heavily on the actions of governments, businesses, and citizens. The possibilities of missteps with enormous environmental consequences are high and the need for constant vigilance even higher. Over the next decades, we will collectively write the next chapter to this unfolding story.

In part I of the book, Gus Speth calls for "a rapid ecological modernization of industry and agriculture," Lester Brown advocates investment in the "technical infrastructure of a new eco-economy," and Amory and Hunter Lovins describe dramatic advances in "resource productivity" that can make a transition to sustainability possible. The technical developments they describe move us far beyond the old days when environmental protection meant ensuring minimum compliance with pollution control regulations into a new era where far greater improvements in environmental quality can be achieved by accelerating the development and use of more advanced technologies for manufacturing, energy, transportation, and agriculture.

In part II, we take a closer look at the issue of technology and its role in environmental protection and sustainability. Employing new technologies in a way that improves environmental quality and minimizes unintended consequences will present society with major challenges involving the mechanisms of governance, the role of leadership, and our perception of time. Most thinking about a transition to sustainability has focused on a limited set of technologies, such as renewable energy systems. But many other areas of rapid technological change will be extremely important—for better or for worse—in creating a sustainable future. How these technologies are developed and deployed and how we as a society deal with their consequences, both intended and unintended, present major governance challenges.

As Mark Wiesner and Vicki Colvin argue in their article, the near-term applications of nanotechology to membranes, catalysis, contaminant sensing, energy production and storage, and contaminant immobilization could produce order-of-magnitude environmental improvements over the current generation of technologies. But these applications also raise serious questions about the potential toxicity and persistence of nanomaterials and their interaction with other chemical substances and with organisms. Feng Zhao and John Seely Brown explore the emerging world of distributed computation where intelligence is everywhere. How could this ubiquitous information fabric be used for environmental gains?

Gary Marchant discusses how advances in genomics could revolutionize our understanding of how chemicals in our environment impact humans and other species. This new knowledge will also generate hard choices in terms of how to intervene in the environment, who is responsible, and who controls newly emerging information on genetic sensitivities and susceptibilities.

The rapid spread of global manufacturing systems and mobile, modular production units will challenge many of our preconceptions of how things are built and the environmental policies designed for traditional manufacturing. Tim Sturgeon explores how the transformation in production may impact environmental performance at global and local levels. Finally, Brad Allenby confronts us with probably the greatest technological and ethical challenge facing humankind in the twenty-first century, the engineering and management of an extremely large system—the earth itself.

In Part III, we look at changes needed in governance. Traditional policy approaches based on hierarchical systems of command and control and market interventions will need to be complemented by the use of networks to steer change. The environmental movement and other civil society movements have become an influential part of global governance by making use of network forms of organization and strategy. As David Ronfeldt argues in his piece, network dynamics can increasingly enable policymakers, business leaders, and social activists to create new mechanisms for joint consultation and cooperation.

This new world will demand a new type of environmental leader—part poet, part scientist, part moral philosopher, part politician—who can both create a compelling vision of where we need to go and mobilize action to get there. As leadership scholar Joanne Ciulla writes in her article, guiding change wisely requires "transformational leaders" in environmental organizations and government who regard other constituencies and agencies as potential allies, not competitors or enemies, and who are willing to work to bring groups together and find areas of consensus.

Social innovator Steward Brand makes the point that accelerating technological change risks creating a pathologically short attention span among people that makes them oblivious to comparatively "slow problems" like species extinction and climate change. Steady, farsighted governance is needed to set the long-term goals for solving critical but slow problems and to maintain the constancy and patience needed to see them through.

The third section also looks at leverage points and particular areas where stepwise interventions could result in larger social transformations. First, a shift is needed from a strategy that aims at reducing the release of toxic chemicals to one that attempts to eliminate toxic emissions altogether—by design. This means, as William McDonough and Michael Braungart argue, creating supply chains and manufacturing processes modeled on nature's cradle-to-cradle cycles, in which one organism's waste becomes food for another. Small efforts within the EPA such as its Green Chemistry, Design for the Environment, and Product Stewardship programs can be the model for a new, high-leverage relationship between government and commerce in which government encourages innovative, ecologically intelligent industrial design and helps to reinvent our global business strategy.

The piece by David Bell argues that only business can quickly and effectively drive the transformation of technology. A key governance challenge, therefore, is to support private sector efforts in this direction. Virtually all the major roles of government can be marshaled into a comprehensive effort to advance corporate sustainability: visioning and goal-setting; collaborating and partnering; leading by example; creating appropriate framework conditions; introducing ecological fiscal reforms; serving as a technological innovator and catalyst; regulating; organizing voluntary and nonregulatory initiatives; and serving as an educator, persuader, and information provider. Understood properly, sustainability is not simply a topic to be added to the agenda of governments; it is a lens through which to view the entire agenda in order to develop integrated strategies.

This does not mean a homogeneous strategy vis-à-vis business. John Elkington writes that the roles of government need to be different in relation to different kinds of corporations. Government has to take the offensive with regulation and enforcement when dealing with corporations highly destructive of natural and human capital. But very different strategies are needed for companies at various stages along the way to corporate sustainability. A comprehensive approach involves a wide range of stakeholders and coordinates across many areas of government policy, including tax policy, technology policy, economic development policy, labor policy, security policy, and so on.

The pieces by Denis Hayes and Hazel Henderson address fundamental changes that must occur in our global financial and trade systems. The International Monetary Fund, the World Bank, the World Trade Organization, and other parts of the global architecture of finance and trade need to be redirected and made more transparent and accountable. New international institutions are needed, including a World Environment Organization.

We end the book with three scenarios that synthesize the main themes of the book—technology and governance. Through these scenarios, we explore a future in which transformational technologies combine with transformed governance to create a more sustainable planet. Other futures are also clearly possible, in which old governance unsuccessfully confronts rapid technological change and our institutions play "catch up" to a rapidly evolving set of technology-induced problems or government actually hinders positive environmental developments. By focusing on technology and governance, we are not implying that these issues alone will predetermine our environmental future, but we believe that, at this point in time, these variables are critical.

In these possible futures, we find ourselves moving much faster, propelled by a continuing flow of innovation and scientific advance. The one thing that we have less of is time—time to react, time to shape, and time to learn. In fact, the speed of technological development forces us to change the way we learn. In the past, we have usually "learned too late" about the negative environmental impacts of new technologies, so that government responses did not occur until the new technologies were widespread and impacts on the environment and human health were already high and, in some cases, irreversible. The greatest environmental protection challenge is to build a capacity for fast anticipatory learning that can identify and head off potential environmental problems in the early stages of developing new technologies.

As we stand on the cusp of the new millennium, we have another chance, a chance to create a sustainable future. We can continue to squander our natural resources, but the real waste would be to squander a historically unique opportunity to shape our emerging technologies, economies, and governance structures for the betterment of the planet. To grasp this chance, we need to move beyond business-as-usual toward more anticipatory approaches to environmental protection and more comprehensive strategies for promoting the transformation of technology. The management guru Peter Drucker often made the point that the theory and practice of business is only a hypothesis and, as a hypothesis, needs to be continually tested. An apt analogy can be made with our environmental and natural resource policies. We can continue to throw old solutions at new problems, or we can experiment and evolve. Our governments, businesses, and citizens need to realize that on the grand and unfolding stage of human evolution the most valued capacity will be our ability to anticipate and shape the emerging set of conditions that will determine our destiny.

Part I

The Goal: A Transition to Sustainability

Perhaps there is a kind of silver lining to these global environmental problems, because they are forcing us, willy-nilly, no matter how reluctant we may be, into a new kind of thinking. . . . Out of the environmental crises of our time should come, unless we are much more foolish than I think we are . . . a redirection of technology to the benefit of everyone, a binding up of the nations and generations, and the end of our long childhood.

—Carl Sagan, 1991

1

Creating a Sustainable Future: Are We Running out of Time?

James Gustave Speth

After hearing hours of scientific testimony on the Clean Air Act, Senator Ed Muskie once asked with frustration, "Aren't there any one-armed scientists?" The panel looked perplexed. Muskie continued, "We've had too much of 'on the one hand, on the other hand!'"

I'm afraid my assessment of the environmental challenges and opportunities ahead will have a little of that two-armed flavor. I want to begin by reviewing several disturbing trends and conclude on a hopeful note, reviewing some recent developments that are indeed very encouraging.[1]

Disturbing Trends

The gravity of emerging global-scale environmental problems was communicated clearly to policymakers in the *Global 2000 Report* to the president at the end of the Carter administration. Other reports—from the United Nations Environment Programme, the Worldwatch Institute, the National Academy of Sciences, and elsewhere—were saying much the same around this time. *Global 2000* got some things wrong, but on the big issues like population growth, species extinction, deforestation, desertification, and global warming its projections of what would happen if societies did not take corrective action have turned out to be all too accurate.

In other words, our political leaders were on notice twenty years ago that there was a new environmental agenda, more global, more threatening, and more difficult than the one that spurred the environmental awakening of the late 1960s and early 1970s. Today, our information on global environmental trends is far more complete and sophisticated, but it is not more reassuring. Here are a few examples:

- Half the tropical forests are gone, and non–Organisation for Economic Co-operation and Development (OECD) countries are projected to lose another 15 percent of their forests by 2020. But this data gives an unduly rosy picture. The cumulative impacts of fire, El Niño–driven drought, and fragmentation in major forest areas such Brazil and Borneo exacerbate the effects of deforestation. And much of what's left is under contract for logging.
- A quarter of all bird species are extinct, and another 12 percent are listed as threatened. Also threatened are 24 percent of mammals, 25 percent of reptiles and amphibians, and 30 percent of fish species. The rate of extinctions today is estimated at one hundred to one thousand times the background rate.
- We are now appropriating about 40 percent of nature's net photosynthetic product annually. We are consuming half the available fresh water. Most people will soon live in water-stressed areas. We are fixing nitrogen at rates that far exceed natural rates and among the many consequences of the resulting overfertilization are fifty dead zones in the oceans, including one in the Gulf of Mexico that is the size of New Jersey.
- In 1960, 5 percent of marine fisheries were either fished to capacity or overfished. Today 70 percent of marine fisheries are in this condition.
- Hardest hit of all are freshwater ecosystems around the globe.
- Over these chilling descriptions of biotic impoverishment looms the biggest threat of all—global climate change. Few Americans appreciate how close we are to the widespread devastation of the American

landscape. The best current estimate is that climate change will make it impossible for about half the American land to sustain the types of plants and animals now on that land. A huge portion of our protected areas—everything from wooded lands held by community conservancies, to national parks, forests, and wilderness—is now threatened. In one projection, the much-loved maple-beech-birch forests of New England will simply disappear. In another, much of the Southeast will become a huge grassland savannah unable to support forests because it will be too hot and dry.

We know what is driving these global trends. There has been more population growth in the few decades since astronauts first walked on the moon than occurred across all the millennia before. It took all of history for the world economy to grow to $6 trillion in 1950. Today, it grows by more than that every five to ten years.

Looking ahead, the world economy is poised to double and then double again in the lifetimes of today's students. We could not stop this growth if we wanted to, and most of us would not stop it if we could. Half the world's people live on less than two dollars per day. They both need and deserve something better. Economic expansion at least offers the potential for better lives, though its benefits in recent decades have been highly skewed.

The OECD estimates that its members' CO_2 emissions will go up by 33 percent between 1995 and 2020. Motor vehicle miles traveled in OECD countries are expected to rise by 40 percent by 2020. The U.S. Energy Information Agency predicts a 62 percent increase in global CO_2 emissions over the same period.

The implications of all this are profound. Let me put it this way: we are entering the endgame in our relationship with the natural world. The current Nature Conservancy campaign has an appropriate name: they are seeking to protect the Last Great Places. We are in a rush to the finish. Soon, metaphorically speaking, whatever is not protected will be paved.

We dominate the planet today as never before. We have a tremendous impact on its great life support systems. Nature as something before and beyond us is gone. We are in a radically new moral position because we are at the planetary controls.

Looking back, it cannot be said that my generation did nothing in response to *Global 2000* and similar alerts. The two basic things we've done are research and negotiate. The scientific outpouring of these twenty years has been remarkable and framework conventions have been established on climate, desertification, and biodiversity, to mention the most notable ones.

The problem is that these framework conventions do not compel action. In general, international environmental law suffers from vague agreements, poor enforcement, and understaffing. We still have a long, long way to go to make these treaties effective. A deeper question is whether we are even on the right track with the recent emphasis on the convention and treaty approach. Were we, mesmerized by the Montreal Protocol, launched on the wrong track altogether?

It would be comforting to think that we have spent these twenty years getting ready and are now prepared to act—comforting but wrong, as is readily apparent from President Bush's abandonment of the Kyoto Protocol. Unfortunately, the political leadership today does not care about these issues and the public does not seem to remember much that we learned in the 1970s.

In sum, the problems are moving from bad to worse, we are unprepared to deal with them, and we presently lack the leadership to even get prepared.

Hopeful Developments

Now, on the other hand, I want to sketch seven transitions needed for the overall shift to sustainability and ask whether there are signs of hope in each area. I believe that there are.

Demographics

The first is the need for an early demographic transition to a stable world population. Here there is definite progress. The midrange projection for 2050 was recently 10 billion people, now it is 9 billion. Analyses suggest an escalation of proven approaches could reduce this number to 7.3 billion, with global population leveling off at 8.5 billion. The main need here is adequate funding for the internationally agreed-upon Cairo Plan of Action.

Human Development

The second transition is the human development transition to a world without mass poverty, a world of greater economic and social equity. We need this transition not only because, over much of the world, poverty is a destroyer of the environment, but also because the only world that will work is one in which the aspirations of poor people and poor nations for fairness and justice are being realized. The views of developing countries in international negotiations on the environment are powerfully influenced by their underdevelopment, concern about the high costs of compliance, and distrust of the intentions of already industrialized countries. Sustained and sustainable human development provides the only context in which there can be enough confidence, trust, and hope to ground the difficult measures needed to realize environmental objectives.

There is good news to report on the human development front. Since 1960, life expectancy in developing regions has increased from forty-six years to sixty-two. Child death rates have fallen by more than half. Adult literacy rose from 48 percent in 1970 to 72 percent in 1997. The share of people enjoying at least medium human development in the United Nations Development Programme (UNDP) Human Development Index rose from 55 percent in 1975 to 66 percent in 1997.

On the policy front, a wonderful thing has happened. The international development assistance community has come together with a concerted commitment to the goal of halving the incidence of absolute poverty by 2015, and all governments in the Millennium Assembly of the United Nations have endorsed this goal. Eliminating large-scale poverty is not a crazy dream. It is within our reach. However, as with population, a principal threat to achieving the goal is declining development assistance.

Technology

The third transition is a transformation in technology to a new generation of environmentally benign technologies—to technologies that sharply reduce the consumption of natural resources and the generation of residual products per unit of prosperity.

We need a worldwide environmental revolution in technology—a rapid ecological modernization of industry and agriculture. The prescription is straightforward but immensely challenging: the only way to reduce pollution and resource consumption while achieving economic growth is to bring about a wholesale transformation in the technologies that today dominate manufacturing, energy, transportation, and agriculture. We must rapidly abandon the twentieth-century technologies that have contributed so abundantly to today's problems and replace them with more advanced twenty-first-century technologies designed with environmental sustainability in mind.

The good news here is that across a wide front sustainable technologies are either available or soon will be. From 1990 to 1998, when oil and natural gas use grew at a rate of 2 percent annually, and coal consumption grew not at all, wind energy grew at an annual rate of 22 percent and photovoltaics at 16 percent. I use an energy example because transformation of the energy sector must rank as the highest priority.

Consumption

The fourth transition is a transition in consumption from unsustainable patterns to sustainable ones. Here one very hopeful sign is the emergence of product certification and green labeling and public support for it. This trend started with the certification of wood products as having been produced in sustainably managed forests and has now spread to fisheries. Many consumers care, and that is driving change.

Markets

The fifth transition is a market transition to a world in which we harness market forces and in which prices reflect environmental costs. The revolutions in technology and consumption patterns just discussed will not happen unless there is a parallel revolution in pricing. The corrective most needed now is environmentally honest prices. Doing the right thing environmentally should be cheaper, not more expensive, as it so often is today.

Here one of the most hopeful developments is the tax shift idea adopted in Germany. Moving in four stages, starting in 1999, the policy is to shift the tax burden from something one wants to encourage—work and the wages that result—to something one wants to discourage—fossil fuel consumption and the pollution that results.

Governance

The sixth transition is a transition in governance to responsible, accountable governments and to new institutional arrangements, public and private, that focus energies on the transition to sustainability. UNDP estimates that today about 70 percent of the people in the developing world live under relatively pluralistic and democratic regimes. Progress on this front is sine qua non.

At the international level, there are governance regimes that have worked: the Montreal Protocol for protecting the ozone layer, the Convention on International Trade in Endangered Species (CITES) for regulating trade in endangered species, International Convention for the Prevention of Pollution from Ships (MARPOL) for pollution from ships. International regulatory processes can be made to work.

And at the local level there is a remarkable outpouring of initiatives: the smart growth movement, sustainable communities and the "new urbanism," state and local greenplans, environmental design in buildings, and innovative state regulatory approaches.

The certification movement mentioned above is an example of still another pathbreaking phenomenon: the rise of information-rich nonregulatory governance, even nongovernmental governance. The forest certification movement is occurring with governments watching from the sidelines. A long list of techniques: the U.S. Toxics Release Inventory and other "right to know" disclosures, third-party auditing, and market creation by government entities and consumers—all coupled with the

Internet and an increasingly sophisticated international nongovernmental organization (NGO) community—can make a powerful contribution.

Meanwhile, in the area of corporate governance and leadership, we are seeing some extraordinary developments:

- Seven large companies—DuPont, Shell, BP Amoco, and Alcan among them—have agreed to reduce their CO_2 emissions 15 percent below their 1990 levels by 2010.
- Today, more than $2 trillion reside in socially and environmentally screened funds. The number of screened mutual funds has grown dramatically in recent years.
- Eleven major companies—DuPont, General Motors, and IBM among them—have formed the Green Power Market Development Group and committed to develop markets for 1,000 megawatts of renewable energy over the next decade.
- Home Depot, Lowes, Andersen, and others have agreed to sell wood (to the degree it's available) only from sustainably managed forests certified by an independent group against rigorous criteria. Unilever, the largest processor of fish in the world, has agreed to the same regarding fish products.

These are among the most hopeful, optimism-generating things I've seen lately.

We are thus far beyond the old days of environment as pollution control compliance. Environment is becoming central to business strategic planning. Companies are beginning to develop sustainable enterprise strategies that are leading to new processes and products and new profits. The war between business and environment should be over, with both sides winning.

Human Culture and Consciousness

Finally, there is the most fundamental transition of all—a transition in culture and consciousness. The potential is evident in great social movements that societies have already experienced, such as the abolition of slavery and the civil rights movement. It seems to me at least possible that we are seeing the beginning of another historic change of consciousness: in the young people on the streets of Seattle, in the far-reaching and unprecedented initiatives being taken by some private corporations, in the growth of NGOs and their innovations, in scientists speaking up and speaking out, and in the increasing prominence of religious and spiritual leaders in environmental affairs.

These are all hopeful signs, but to be honest we must conclude that we are at the early stages of the journey to sustainability. Meanwhile, the forward momentum of the drivers of environmental deterioration is great. We are moving rapidly to a swift, pervasive, and appalling deterioration of our natural world. Time is the most important variable in the equation of the future. What we will do tomorrow we should have done yesterday. Only a response that in historical terms would come to be seen as revolutionary is likely to avert these changes.

For Further Exploration

Yale School of Forestry and Environmental Studies web site: www.yale.edu/forestry/.

Notes

1. This essay draws upon Dean Speth's book, *Red Sky at Morning: America and the Crisis of the Global Environment* (New Haven: Yale University Press, 2004).

2

An Eco-Economy in Harmony with Nature

*Lester Brown**

In 1543, Copernicus published a paper, "On the Movement of the Celestial Spheres," in which he put forth the radical idea that the sun does not revolve around the earth; rather, the earth revolves around the sun. Copernicus's revolutionary idea inaugurated debate between scientists and theologians that lasted for centuries. His new perspective set the stage for enormous progress in astronomy and physics and in all of the related sciences.

Today we are in a somewhat similar situation. The question is not whether the sun revolves around the earth or the earth around the sun, but whether the economy is part of the environment or the environment is part of the economy. Most economists, and I think it would be fair to say most members of the business community, think of the environment as being a subset of the economy. It is the pollution sector.

Ecologists, on the other hand, see the economy as part of the earth's ecosystem. It is the commercialized part. If the ecologists are right, then

*Excerpted from Lester R. Brown, *Eco-Economy: Building an Economy for the Earth* (W. W. Norton & Co., NY: 2001).

it follows that the economy must be designed so that it is compatible with the earth's ecosystem.

To operate in harmony with nature, our local, national, and global economies need to respect the principles of ecology. These principles are as real as those of aerodynamics. If an aircraft is to fly, it has to satisfy certain principles of thrust and lift. So, too, if an economy is to sustain progress, it must satisfy basic principles of ecology. If it does not, it will decline and eventually collapse. There is no middle ground. An economy is either sustainable or it is not.

Out of Balance

Our existing economy is not sustainable. It is out of sync with the earth's ecosystem; it is destroying its own natural support systems. Over the last half century, a sevenfold expansion of the global economy has pushed the demand on local ecosystems beyond the sustainable yield in country after country. The fivefold growth in the world fish catch since 1950 has pushed most oceanic fisheries past their ability to produce fish sustainably. The sixfold growth in the worldwide demand for paper is shrinking the world's forests. The doubling of the world's herds of cattle and flocks of sheep and goats since 1950 is damaging rangelands, converting them to desert.

The market in and of itself does not recognize basic ecological concepts or respect the balances of nature. And in a world where the demands of the economy are pressing against the limits of natural systems, basing investments on price signals that carry no information about environmental costs is a recipe for disaster. Historically, for example, when the supply of fish was inadequate, the price would rise, encouraging investment in additional fishing trawlers. When there were more fish in the sea than we could ever hope to catch, the market worked well. Today, with the fish catch often exceeding the sustainable yield, investing in more trawlers in response to higher prices will simply accelerate the collapse of these fisheries.

A similar situation exists with other natural systems such as aquifers, forests, and rangelands. Once the climbing demand for water surpasses the sustainable yield of aquifers, the water tables begin to fall and wells go dry. The market says drill deeper wells. Farmers engage in a competitive orgy of well drilling, chasing the water table downward. On the North China Plain, where a quarter of the country's grain is produced, this process is well underway. In Hebei Province, data for 1999 show thirty-six thousand mostly shallower wells having being abandoned as fifty-five thousand new, much deeper wells were drilled. In Shandong Province, thirty-one thousand were abandoned and sixty-eight thousand new wells were drilled.

Creating an Eco-Economy

An eco-economy would be one that satisfies our needs without jeopardizing the prospects of future generations to meet their needs. It would respect the sustainable yield of the ecosystems on which it depends: fisheries, forests, rangelands, and croplands. It would respect the balances maintained by natural systems. These include balances between soil erosion and new soil formation, between carbon emissions and carbon fixation, and between trees dying and trees regenerating. It would pattern itself on nature's cyclical processes, with no linear flow-throughs, no situations where raw materials go in one end and garbage comes out the other. In nature, one organism's waste is another's sustenance, and nutrients are continuously cycled. This system works. Our challenge is to emulate it in the design of the economy.

Converting our economy into an eco-economy is a monumental undertaking. There is no precedent for transforming an economy shaped largely by market forces into one where markets operate within the framework of principles of ecology. Yet partial glimpses of the eco-economy are already visible in many countries.

For example, thirty-one countries in Europe, plus Japan, have stabilized their population size, satisfying one of the most basic conditions of an eco-economy. A reforestation program in South Korea, begun more than a generation ago, has blanketed the country's hills and mountains with trees. Costa Rica has a plan to shift entirely to renewable energy by 2025. Iceland, working with a consortium of corporations led by Shell and DaimlerChrysler, plans to be the world's first hydrogen-powered economy. Denmark is emerging as the eco-economy leader: it has stabilized its population, banned the construction of coal-fired power plants, banned the use of nonrefillable beverage containers, restructured its urban transport network, and is now getting 15 percent of its electricity from wind.

Restructuring the Economy

Describing the eco-economy is obviously a somewhat speculative undertaking. In the end, however, it is not as open-ended as it might seem, because the eco-economy's broad outlines are defined by the principles of ecology. We can already foresee many of its major features.

- Energy and materials of all kinds will be used far more efficiently than they are today.
- The energy system will be hydrogen-based rather than carbon-based. Instead of running on fossil fuels, it will be powered by renewable sources of energy such as wind and sunlight, and by geothermal energy from within the earth. Cars and buses will be powered by fuel cells that use hydrogen as a fuel and have no emissions aside from pure water.
- Atmospheric carbon dioxide levels will be stabilized. In contrast to today's energy economy, where the world's reserves of oil and coal are concentrated in a handful of countries, energy sources in the eco-economy will be as widely dispersed as sunlight and wind. The world economy will no longer depend on Middle Eastern oil.
- Transport systems will become more diverse. Cars will still be available as needed, but instead of the congested, polluting, auto-centered transport systems of today, urban areas will have more rail-centered transport systems and will be more bicycle and pedestrian friendly, offering easier access, more exercise, cleaner air, and less frustration.
- Materials use will shift from the linear economic model, where materials go from the mine or forest to the landfill, to the reuse/recycle model, which yields no waste for the landfills.
- The use of water will be in balance with supply. Water tables will be stable, not falling. Water-efficient technologies will raise water productivity in every facet of economic activity.

- Renewable resource use will be kept below the maximum sustainable yield. For example, harvests from oceanic fisheries, a major source of animal protein in the human diet, will be reduced to the sustainable yield and fish farming will expand to satisfy additional demand. The excessive pressure on rangelands will be alleviated by measures such as feeding livestock crop residues that are otherwise wasted or burned for fuel.
- And finally, the new economy will be based on a stable population. Over the longer term, the only sustainable society is one in which couples have an average of two children.

Opportunities in the New Economy

The real "new economy" is still ahead. Building that new economy involves phasing out old industries, restructuring existing ones, and creating new ones. The world coal industry is an example of an industry already being phased out, dropping 7 percent since peaking in 1996. This decline will continue in an eco-economy unless we find cost-effective ways to extract hydrogen from coal and sequester carbon dioxide.

The automobile industry faces a major restructuring as it changes power sources, shifting from the gasoline-powered internal combustion engine to the hydrogen-powered fuel-cell engine. This shift will require both a retooling of engine plants and the retraining of automotive engineers and automobile mechanics.

The new economy will also create major new industries, ones that either do not yet exist or that are just beginning. Wind is one such industry. Now in its embryonic stage, it promises to become the foundation of the new energy economy. Millions of turbines soon will be converting wind into electricity, becoming part of the global landscape.

As wind power emerges as a low-cost source of electricity and a mainstream energy source, it will spawn another industry: hydrogen production. Once wind turbines are in wide use, there will be a large unused capacity during the night when the demand for electricity drops. With this essentially free electricity, turbine owners can turn on the hydrogen generators and convert wind power into hydrogen. Hydrogen generators will start to replace oil refineries.

Changes in the world food economy will also be substantial. Some of these, such as the shift to fish farming, are already underway. The fastest growing subsector of the world food economy during the 1990s was aquaculture, expanding by more than 11 percent a year. Fish farming is likely to continue to expand simply because of its efficiency in converting grain into animal protein. Even allowing for slower future growth in aquaculture, fish farm output will likely overtake beef production before 2010. Perhaps more surprising, fish farming could eventually exceed the oceanic fish catch.

Just as the last half century has been devoted to raising land productivity, the next half century will be focused on another growth industry: raising water productivity. Virtually all societies will turn to the management of water at the watershed level in order to manage the available supply most efficiently. Irrigation technologies will become more efficient. Urban wastewater recycling will become common. At present, water tends to flow into and out of cities, carrying waste with it. In the future, water will be used over and over, never discharged. Since water does not wear out, there is no limit to how long it can be used, as long as it is purified before reuse.

Teleconferencing is another industry that will play a prominent role in the new economy and reduce energy use. Increasingly for environmental reasons and to save time, individuals will be "attending" conferences electronically with both audio and visual connections. This industry involves developing the global electronic infrastructure, as well as the services, to make teleconferencing possible. One day there may be thousands of firms organizing electronic conferences.

New Jobs in the Eco-Economy

Restructuring the global economy and creating new industries will create new jobs—indeed, whole new professions and new specialties within professions. For example, as wind becomes an increasingly prominent energy source, tens of thousands of new jobs will be created in turbine manufacturing, installation, and maintenance, as well as the thousands of wind meteorologists who will be needed to select the best sites for wind farms.

Environmental architecture will be another fast-growing profession. In a future of water scarcity, watershed hydrologists will be in high demand. As the world shifts away from a throwaway economy, engineers will be needed to design products that can be disassembled quickly and easily into component parts and materials and reused or recycled. Recycling engineers will be responsible for closing the materials loop, converting the linear flow-through economy into an "industrial ecosystem" based on comprehensive recycling.

If the world is to stabilize population sooner rather than later, it will need far more family-planning midwives in Third World communities. This growth sector will be concentrated largely in developing countries, where millions of women lack access to family planning. The same family-planning counselors who advise on reproductive health and contraceptive use can also play a central role in mobilizing their societies to control the spread of HIV.

Another pressing need, particularly in developing countries, is for sanitation-system engineers who can design sewage systems not dependent on water, a trend already underway in some water-scarce countries. As it becomes clear that using water to wash waste away is a reckless use of a scarce resource, a new breed of sanitation engineers will be in wide demand.

Investing in the Environmental Revolution

There has never been an investment situation like this before. The amount that the world spends now each year on oil, the leading source of energy, provides some insight into how much it could spend on energy in the eco-economy. In 2000, the world used nearly 28 billion barrels of oil, some 76 million barrels per day. At $27 a barrel, the total comes to $756 billion per year. How many wind turbines, solar rooftops, and geothermal wells will it take to produce this much energy?

Investments in a new energy infrastructure will be huge. They include the transmission lines to connect wind farms with electricity consumers and the pipelines to link hydrogen supply sources with end users. For developing countries, the new energy sources promise to reduce dependence on imported oil, freeing up capital for investment in domestic energy sources. Investments in energy efficiency are also likely to grow rapidly simply because they are so profitable. In virtually all countries, industrial and developing, saved energy is the cheapest source of new energy. Replacing inefficient incandescent light bulbs with highly efficient compact fluorescent lamps offers a rate of return that stock markets cannot match.

Investment opportunities will abound in the food economy. It is likely that the world demand for seafood, for example, will increase at least by half over the next fifty years, and perhaps much more. If so, fish-farming output—now 31 million tons a year—will roughly need to triple, as will investments in fish farming. Although aquaculture is likely to slow from its 11 percent yearly growth of the last decade, it is nonetheless likely to be robust, presenting a promising opportunity for investors.

A similar situation exists for tree plantations. At present, tree plantations cover some 113 million hectares (280 million acres). An expansion of these by at least half, along with a continuing rise in productivity, is likely to be needed both to satisfy future demand and to eliminate one of the pressures on shrinking forests. This, too, presents a huge opportunity for investment.

Accelerating the Transition

Making the transition to an eco-economy is the only way that economic progress can be sustained. The longer old industries and their political allies succeed in delaying this transition, the more disruptive it will be. If many nations delay too long, they may undermine the ecological foundations on which their economies are built. To avoid these dangers, we need to reach agreement as rapidly as possible on the need for systemic change. We will not succeed with marginal improvements in a few environmental regulations here and a few new projects there.

Governments need to explicitly take on the challenge of formulating clear goals and coherent strategies to put the world on an environmentally sustainable development path. Any lesser role is not enough.

The private sector, however, will be where the eco-economy is primarily built. No sector of the global economy will be untouched. In this new economy, some companies will be winners and some will be losers. Those who anticipate the emerging eco-economy and plan for it will be the winners. Those who cling to the past risk becoming part of it.

For Further Exploration

Lester Brown, *Eco-Economy: Building an Economy for the Earth*. New York: W. W. Norton, 2001.

Lester Brown, Plan B: Rescuing a Planet under Stress and a Civilization in Trouble. New York: W. W. Norton, 2004.

Lester Brown et al., *The Earth Policy Reader*. New York: W. W. Norton, 2002.

Earth Policy Institute web site: www.earth-policy.org/.

3

A New Age of Resource Productivity

Amory B. Lovins and L. Hunter Lovins

The transition to sustainability will require a new industrial revolution based on a high level of scientific knowledge and technological sophistication. This conclusion follows from the growing realization that the environmental problem is not so much a polluted river here or a release of a particular toxin, but the worldwide loss of the ecosystem services that underpin all life and thus all economic activity. Environmental protection, as it is usually conceived, cannot solve this problem. What is required is a modernization of our entire technological infrastructure to eliminate waste and pollution by radically improving resource productivity and creating closed-loop industrial systems that mimic biological processes.

The First Industrial Revolution: Increasing Labor Productivity

The first industrial revolution grew out of conditions in which the scarcity of skilled labor was limiting material progress. Before that time, it was inconceivable that people could work much more productively. If you wanted more cloth, you had to hire more skilled weavers—if you could find them. So it made sense to use machines, energy, and resources to allow each worker to produce more.

The textile mills introduced in the late 1700s soon enabled one Lancashire spinner to produce the cloth that had previously required two hundred weavers. As many such technical and organizational inventions improved the productivity of workers in sector after sector of the economy, affordable mass goods, purchasing power, a middle class, and everything we now call the industrial revolution emerged. All of our economic arrangements today, from tax codes to mental models, derive from this effort to economize on the scarcest factor of production, skilled people, and substitute the use of seemingly abundant nature to supply resources and absorb pollution.

The logic of economizing on the scarcest resource—because that is what limits progress—remains perennially true. What has changed—indeed, reversed—is the pattern of scarcity. Today we have abundant people and scarce nature, not the other way around. This is not to say that commodities are scarce. What is increasingly limited is the ability of deteriorating living systems to provide the *ecosystem services* needed to sustain growing populations and economies.

Ecosystem services are the natural processes that cycle nutrients and water, regulate the atmosphere and climate, provide pollination and biodiversity, rebuild topsoil and biological productivity, control pests and diseases, and assimilate and detoxify society's wastes. These free and automatic services provide tens of trillions of dollars of worth each year—more than the entire global economy. Indeed, their value is nearly infinite, since without them there is no life and therefore no economic activity. Yet none of their value is reflected on anyone's balance sheets. As a result, ecosystem services are diminishing. As the recent report by the United Nations, the World Bank, and the World Resources Institute, *People and Ecosystems: The Fraying Web of Life*, puts it, "There are considerable signs that the capacity of ecosystems, the biological engines of the planet, to produce many of the goods and services we depend on is rapidly declining."

Traditional environmental protection measures cannot by themselves reverse this decline. A completely different approach is needed. Today's patterns of relative scarcity and abundance dictate using more people and more brains to wring four, ten, or even one hundred times more benefit from each unit of energy, water, materials, or anything else borrowed from the planet. This shift in relative scarcities is already beginning to move the market. Forward-looking firms seek not just greater labor productivity, but total factor productivity that uses all resources more efficiently. Increased resource productivity will be the hallmark of what Paul Hawken calls the "Next Industrial Revolution."

The Next Industrial Revolution: Increasing Resource Productivity

Dramatic improvements in resource productivity are relatively easy to achieve because resources of all kinds are used incredibly wastefully now. The stuff that drives the metabolism of industry currently amounts to more than twenty times your body weight every day, or more than 1 million pounds per American per year. The corresponding figures for Europe and Japan are not that different. Globally, the economy mobilizes a flow of half a trillion tons per year. But only 1 percent of that huge flow ever gets embodied in a product and is still there six months after sale. The other 99 percent is waste.

Reducing that waste represents a vast business opportunity. Shifts already underway toward lean manufacturing systems and water-efficient technologies for agriculture, industry, and buildings are occurring because they cut costs and boost profits as well as slash environmental impacts. Nowhere are opportunities of this kind easier to see than in energy.

By using energy more efficiently, Americans cut oil use 15 percent in the six years after the 1979 oil shock while the economy grew 16 percent. Since then, more efficient use has grown to become America's biggest energy "source"—not oil, gas, coal, or nuclear power. There are many ways to measure progress in doing more with less energy, but even by the broadest and crudest measure—lower primary energy consumption per dollar of real gross domestic product—progress has been dramatic. By 2000, improved efficiency (compared with 1975) was providing 40 percent of all U.S. energy services (heating and cooling, mobility, and so on). It was 73 percent greater than U.S. oil consumption, five times domestic oil production, three times total oil imports, and thirteen times Persian Gulf oil imports. Growing efficiency is the most important energy development of the past generation, but it has gone largely unnoticed because it hasn't cost a lot, produced environmental problems, posed risks to national security, or called attention to itself in other headline-grabbing ways.

This progress was mostly achieved by more efficient use of energy, partly by shifts in the economic mix, and only slightly by behavioral change. Since 1996, saved energy has been the nation's fastest-growing major "source."

The potential for further improvements is enormous. State-of-the-shelf technologies can make old buildings three- to fourfold more energy efficient, new ones nearer tenfold—and cheaper to build. For example:

- At the Rocky Mountain Institute, high in the Rocky Mountains, efficiency improvements saved 99 percent in space- and water-heating energy, cut electricity use by 90 percent, and paid for themselves in the first ten months—all with 1983 technologies. The building cost less than normal to build because the superwindows, superinsulation, and ventilation heat recovery that let us eliminate the furnace cost less than the furnace would have cost to install.
- Architecture professor Suntoorn Boonyatikarn built a delightful house in tropical Bangkok that uses only 10 percent the normal amount of air-conditioning, yet maintains superior comfort and cost nothing extra to build.
- An existing California office building was cost-effectively improved to save more than 90 percent of its air-conditioning energy while improving comfort.

Industries can achieve similarly surprising savings:

- Southwire Corporation, an energy-intensive maker of cable, rod, and wire, halved its energy use per pound of product in six years. The savings roughly equaled the company's profits during a period when many competitors were going bankrupt. The company then went on to save even more energy, still with two-year paybacks.
- Dow Chemical's Louisiana Division implemented more than nine hundred worker-suggested energy-saving projects from 1981 through

1993, with average annual returns on investment in excess of 200 percent. Both returns and savings tended to rise in the latter years, even after the annual savings had surpassed $100 million, because the engineers were learning new ways to save faster than they were using up the old ones.

- A combination of efficiency improvements can save about half the energy in typical existing industrial motor systems (which use three-fourths of industrial electricity) with returns on investment approaching 200 percent per year.
- In a typical industrial pumping loop, an improved design cut power use by 92 percent, cost less to build, and worked better. This was achieved not by any new technology but solely by better design that used fat, short, straight pipes rather than skinny, long, crooked ones. It wasn't rocket science—just good Victorian engineering rediscovered. But it was important because pumping is the biggest user of electricity worldwide.

The efficiency revolution's latest surprise squarely targets oil's main users and its dominant growth market: cars and light trucks. New American cars average twenty-four miles per gallon (mpg), a twenty-year low. But an industrywide transition is on the horizon. Toyota's Prius hybrid-electric five-seater gets up to sixty mpg. A car fleet as efficient as the Prius would save twenty-five Arctic Refuges, but it's just the start.

In 2000, Hypercar, Inc., designed a competitively priced concept sport utility vehicle (SUV) as roomy, comfortable, and sporty as a Lexus RX-300—and as safe even if hit by one, although the Lexus is twice its weight. The car's structure is made of ultralight carbon-fiber composite, which can absorb up to five times more crash energy per pound than steel. Getting the equivalent of ninety-nine mpg, it would drive 330 miles on 7.5 pounds of safely stored compressed hydrogen. Driving at fifty-five miles per hour the Hypercar would use as much power as a normal SUV needs for its air conditioner.

Cars using the kind of technologies pioneered in the Hypercar design can transform the world's trillion-dollar auto industry within a few decades. Policy interventions to spur people to buy sluggish or unsafe cars won't be needed to save fuel and reduce emissions: the new cars will sell simply because they're better than current models in every way. Fuel-cell electric vehicles have far fewer parts—no internal combustion engine components, no drive train, no conventional hydraulic and mechanical systems. As a result, manufacturers should enjoy a competitive advantage because their needs for capital, parts, space, and assembly could be as much as ten times lower.

Still further efficiency improvements are possible through advances in the way we produce energy. Smaller power sources located at or near the customer, collectively called "distributed generation," offer a number of efficiencies not provided by big, centralized plants. And a shift from hydrocarbons to pure hydrogen will allow widespread distributed generation using fuel cells, the most efficient, clean, and reliable known source of electricity. Initially, the hydrogen will be made mainly from natural gas. In the long run, hydrogen will most likely be made from water, using renewable electricity or possibly just sunlight. Or it may even be extracted from coal without releasing the carbon into the air. All these options are evolving rapidly and will compete vigorously.

The enormous potential for saving energy means that the production of greenhouse gases can be lowered at a profit, because saving fuel costs less than buying fuel. And the shift to a hydrogen economy can stop climate change altogether. This isn't science fiction, it's beginning to happen. DuPont recently announced that by 2010 it will reduce its CO_2 emissions by 65 percent from 1990 levels, raise its revenues 6 percent a year with no increase in energy use, and get a tenth of its energy and a quarter of its raw materials from renewables—all in the name of increasing shareholder value. STMicroelectronics, the world's sixth-largest chipmaker, has set a goal of zero net carbon emissions by 2010 despite a fortyfold increase in production from 1990, again in pursuit of commercial advantage. The heads of seven major oil and car companies have announced the start of both the Oil Endgame and the Hydrogen Era—a future in which they are strongly investing.

Adopting Biological Patterns and Processes

Resource productivity is the cornerstone of the next industrial revolution, but is only its beginning. Beyond reducing waste through improvements in efficiency lies the challenge of eliminating the entire concept of waste by adopting biological patterns and processes.

Adopting the model of natural systems, in which everything is recycled and nothing goes to waste, implies closing the loops in the flow of toxic materials and eliminating any industrial output that represents a disposal cost rather than a saleable product. There should be none of what in the twentieth century were called "wastes and emissions" but are properly called "unsaleable production." If we can't use it and can't sell it, we shouldn't produce it; we should design it out.

DesignTex, a subsidiary of Steelcase (the world's largest manufacturer of office furniture) commissioned architect Bill McDonough and chemist Michael Braungart to develop a green textile for upholstering office chairs. The development team screened more than eight thousand chemicals. They rejected any that were persistently toxic; built up in food chains; or caused cancer, mutations, birth defects, or endocrine disruption. They found only thirty-eight that they were certain weren't harmful. From these, however, they could make every color. The new fabric they developed won design awards. It looked better, felt better in the hand, and lasted longer, because harsh chemicals did not damage the natural fibers. Production also cost less, because it used fewer and cheaper feedstocks and caused no health and safety concerns.

When Swiss environmental inspectors tested the new plant, they thought their equipment was malfunctioning: the water coming out was cleaner than the Swiss drinking water going in, because the cloth itself was acting as an additional filter. But what had really happened was that the redesign of the process had eliminated any waste and toxicity. As architect McDonough put it, the redesign "took the filters out of the pipes and put them where they belong, in the designer's heads."

Nature's cyclical processes provide the model for the kind of closed-loop thinking that will ultimately restructure our industrial technology—and save the chemical industry. Learning to use nature as a model and a mentor will lead to many other exciting technical developments. Some of the most important will be modeled on nature's low-temperature, low-pressure assembly techniques. Spiders make silk as strong as Kevlar but much tougher from digested crickets and flies without needing boiling sulfuric acid and high-pressure extruders. The abalone makes an inner shell twice as tough as ceramics and diatoms make seawater into glass; neither need furnaces. Trees turn sunlight and soil into cellulose, a sugar stiffer and stronger than nylon but much less dense. Then they bind it into wood, a natural composite with higher bending strength than aluminum alloy, concrete, or steel. Yet trees don't need smelters, kilns, or blast furnaces.

The benign natural chemistry of living nanotechnologies is a better model than industrialism's primitive approach of "heat, beat, and treat." It is based on the design experience of nearly 4 billion years of evolutionary testing in which products that didn't work were recalled by the Manufacturer. Though many details of nature-mimicking practices are still being explored, the broad contours of the lessons they teach are already becoming clear. They can be applied in many areas of technology including the development of nonliving nanotechnologies, which in themselves pose profound challenges and concerns.

Solutions and Disruptions

The decades ahead will be a turbulent time as radical improvements in resource efficiency disrupt business-as-usual. Companies that learn to structure their relationships so that both they and their customers make money by finding more efficient solutions will gain a commanding advantage. Those that don't eventually won't be a problem—they won't be around.

In the near term, however, companies that can't or won't adapt will use their political alliances to push for classic protectionist policies such as larger subsidies for fossil fuels, nuclear power, and an array of extractive industries. Fortunately, these efforts will not dominate the policy arena for long because the policy menu for encouraging resource efficiency is so rich and diverse that it can appeal to all ideological tastes. We can eliminate many of the institutional barriers and government policies that prevent the market from dispassionately picking the best portfolio of investments in both resource efficiency and supply. We can teach architecture, engineering, and business students how to make the most of modern efficiency potential. We can make markets in saved energy, so bounty hunters will pursue it relentlessly. We can scrap inefficient technologies as vigorously as we introduce new ones, rather than further impoverishing poor people and poor nations by selling them our cast-off junk.

As the transition to sustainability gains force, it will become politically possible to back away from the tax and subsidy policies that grew out of the first industrial revolution's drive to substitute resources for people. Groups like Redefining Progress are demonstrating how a market-based next industrial revolution strategy of desubsidization and gradual and fair tax shifting can provide more of what we want, jobs and efficiency, and less of what we don't want, waste and environmental damage.

The large and rapid technical changes ahead will force even the best companies, developing the best technologies, to work harder than ever in order to foresee and design out undesirable impacts. This is an area where government-business cooperation can be especially helpful. But the disruptions and difficulties ahead will be worthwhile. The transition to a technically advanced sustainable society will improve manufacturing, housing, mobility, health, the environment, national security, and overall quality of life.

Inventor Edwin Land said that people who seem to have had a new idea have often simply stopped having an old idea. The key old idea to stop having is that progress has to be based on maximizing the flow of resources from the mine and wellhead through the economy to the garbage dump. A far more elegant and sustainable approach is emerging as we enter a new age of resource productivity.

For Further Exploration

Amory B. Lovins, L. Hunter Lovins, and Paul Hawken, *Natural Capitalism*. New York: Little Brown, 1999. See also web site: www.natcap.org.

The Natural Capitalism Group web site: www.natcapgroup.org

Rocky Mountain Institute web site: www.rmi.org.

Part II

New Technologies

He that will not apply new remedies, must expect new evils; for time is the greatest innovator.

—Francis Bacon 1561–1626

4

Environmental Implications of Emerging Nanotechnologies

Mark R. Wiesner and Vicki L. Colvin

Nanotechnology is often described monolithically as an area with great potential for scientific discovery and as a fertile investment sector.[1] Indeed, the vision of building objects from the atomic scale up combines many scientific areas of inquiry with wide implications for technology development. Unlike the more focused biotechnology or information technology (IT) sectors, nanotechnology and the materials behind it encompass multiple sectors (including biotech and IT) and draw on a wide array of substances and formats. The production, use, and disposal of nanomaterials can be anticipated to engender an equally wide range of benefits and unintended consequences in social, economic, and environmental terms.

The breadth and novelty of nanotechnology necessarily require a speculative treatment of any eventual impact. With this caveat, we believe it is very likely that applications of nanotechnology will lead to new means of reducing the production of wastes, using resources more sparingly, remediating industrial contamination, providing potable water, and improving the efficiency of energy production and use. In short, nanomaterials will be essential to achieving many of the goals articulated in the first section of this book.

As nanotechnology emerges as an important force behind these new environmental technologies, we are also presented with an unprecedented opportunity to consider the environmental implications of an emerging technology at its inception. Our ultimate goal is to ensure that nanotechnologies, and the materials that enable them, evolve as instruments of sustainability rather than as environmental liabilities.

We approach this unusual position armed with a multitude of examples from the twentieth century of new technologies that at their start seemed only positive, but in time created environmental problems that detracted from their initial promise. Unfortunately, this transition—from technological optimism to skepticism—has occurred all too often. The environmental community learned a painful lesson in how technology can bite back with the pesticide dichlorodiphenyltrichloroethane (DDT). DDT was credited with a variety of benefits including increased crop yields and effective control of mosquitoes and lice as well as reduced incidence of the diseases these insects carry. Some twenty years later, the environmental consequences of DDT became apparent in the 1960s and were immortalized in Rachel Carson's famous book *Silent Spring*. Today, DDT manufacture in the United States has ceased. But in many parts of the world, where DDT manufacture and use continue, DDT is still seen as a highly beneficial chemical.

The unintended consequences of a technology may be positive as well as negative. In contrast with the problems surrounding the disposal of nuclear wastes associated with nuclear energy production, which were in many cases anticipated and remain largely unresolved today, the benefits of lower carbon dioxide emissions associated with nuclear power generation were not articulated when this technology was conceived.

Today, nanotechnology is at a critical stage where relatively small adjustments may earn large returns in creating a technology with great net environmental benefits. Nevertheless, of the more than $700 million U.S. investment in nanotechnology research in FY2003, less than $1 million was earmarked for studies of overall societal impact, and a similar nominal sum was earmarked for questions directly addressing environmental impact. It is essential that our investments in precautionary research earlier in the trajectory of emerging nanotechnologies are proportionate to the enormous benefits that successful implementation of these technologies may yield.

In this essay, we present an overview of selected areas where we believe that nanotechnology can play an important role in improving environmental quality as well as a consideration of critical areas for research on the possible environmental impacts of a mature nanomaterials industry.

Nanotechnologies as Environmental Technologies

Nanotechnology is likely to produce substantial improvements in environmental technologies and public health in applications such as industrial separations, potable water supply, chemical synthesis, and air-quality control. A key strength of nanotechnology-based approaches to maintaining or improving environmental quality lies in their potential to approach the thermodynamic limits of efficiency for production and cleanup processes, thereby reducing energy consumption per unit of production with associated environmental benefits. Specifically, near-term applications of nanotechnology to the following areas may result in significant improvements in efficiency and reductions in cost:

- Membrane science
- Catalysis
- Contaminant sensing
- Energy production and storage
- Contaminant immobilization

Therefore, it is not unreasonable to anticipate order-of-magnitude improvements over the current generation of technologies. We will examine each of these areas in more detail.

Membrane technologies are playing an increasingly important role for environmental quality control, resource recovery, pollution prevention, energy production, and environmental monitoring.[2] Membranes are also key technologies at the heart of fuel cells and bio-separation devices. Nanoscale control of membrane architecture may yield membranes of greater selectivity and lower cost. Nanotechnology might be used to create smart membranes that integrate sensing and separation capabilities in the same structure, allowing membranes to adaptively select compounds for transport, detect a compromised membrane surface, or adapt to changes in the environment such as temperature or pH.

Catalysis is increasingly replacing the use of many hazardous substances in chemical production to achieve cleaner chemical synthesis.[3] Catalysis is also important in reducing downstream pollutants after they are generated, as is done in the catalytic converters of automobiles. Nanoparticle-based strategies for catalytic remediation of contaminated ground waters have been proposed and tested at field scale.

Numerous strategies have been proposed for developing nanotechnology-based sensors for environmental contaminants. Nanostructured substrates for spectrophotometric measurement[4] have the capability of greatly enhancing the sensitivity of measurements for many environmental contaminants. This will allow for fewer manipulations of samples, direct measurement in the field, and potentially more stringent controls on sources of compounds as our awareness of the presence of contaminants in our air, water, soil, and biosphere increases. Molecular electronics–based devices have the potential of dramatically reducing the size and cost of current analytical devices, with a potential to create distributed analytical networks. Distributed analytical networks would have an impact similar to the development of small inexpensive microprocessors that shifted computing power from centralized facilities to desktops, appliances, and myriad other locations. The capability to perform measurements at many locations will improve both the quality and quantity of environmental information collected and may affect the way sources of pollution are regulated and monitored. Distributed analytical capabilities would also improve the ability to perform real-time process control, thereby improving the efficiency of process with associated reduction in costs and wastes generated.

Energy production and storage are priority areas for nanotechnology research. Improvements in the efficiency and cost of solar cells and batteries are feasible with nanotechnology-based methods for the creation of thin films.[5] The storage of hydrogen, particularly for the development of fuel cells, is an important challenge, which has recently focused on the use of carbon nanostructures. The first reported results for hydrogen storage using carbon single-wall nanotubes (SWNTs) were published in 1997.

Both storage and immobilization of hazardous materials are of great long-term concern for industries ranging from nuclear power generation to chemical production. First-generation solutions involved embedding wastes in solids such as cement or incorporating materials in vitrified masses. Nanotechnology offers the possibility of constructing highly stable masses for long-term storage at the molecular level.

Anticipating Environmental Impacts

Production of significant quantities of anthropogenically derived nanomaterials will inevitably result in the introduction of these materials to the atmosphere, hydrosphere, and biosphere. Key questions to be addressed in future research include those related to elucidating and managing risk such as:

- What will be the environmental impacts associated with producing, using, and disposing of these materials?
- Where will these materials most likely end up in the environment?
- What are the most probable paths of exposure to these materials?
- Are these materials toxic?
- How persistent will these materials be?
- How will these materials interact with other chemicals and with organisms?
- How do the properties and quantities of anthropogenically derived nanomaterials compare with those of naturally derived nanomaterials?

Research is needed to explore the impacts of nanomaterials and nanomaterial production on the environment and public health. One framework for assessing these impacts is that of comparative risk assessment. Applied to an assessment of the production, use, and disposal of nanomaterials, a risk assessment typically considers both the potential for exposure to a given material and (once exposed) potential impacts such as toxicity or mutagenicity. The need to evaluate both of these components of risk in assessing the consequences of nanomaterials on the environment and public health is essential.

Exposure Research

Exposure to nanomaterials will be determined in large part by the chain of production, use, and disposal. Therefore, estimating nanomaterial exposure will probably prove to be a time-consuming process. If nanoparticles were present as aerosols, it might be easier for them to enter the lungs. We note, however, that some nanoparticles, such as carbon nanotubes as they are currently produced, should not easily become aerosols. Direct skin absorption or ingestion may prove to be a more common mode of exposure. Little to nothing is known about nanoparticle exposure through direct skin absorption, ingestion, or intravenous pathways. Some nanomaterials may be introduced intentionally to the environment, as for example the use of nanoparticles to clean up contaminated aquifers. In other cases, these materials may inadvertently enter aquatic environments through product use and disposal.

Despite the need to examine exposure issues for each nanomaterial, mode of production and use, and disposal scenario, research can be conducted in a fashion that will allow the generalization of results. Research should focus on fundamental mechanisms such as those governing transport through biological membranes, transport in air and water, and interactions with environmental surfaces.

Although the mobility of nanoscale particles in the environment is poorly understood, speculation based on laboratory experience in making nanomaterials and experience garnered from colloid science[6] suggests that the mobility of anthropogenic nanoparticles, and thus the potential for exposure, may be small. Nonetheless, the relevance of our current experience in this domain is unclear. Although there are many nanoscale particles produced naturally, it is unclear how man-made materials might differ in their interactions in ground and surface waters. In particular, the impact of nanoparticle chemistry and interactions with surfaces are of great interest. It is also unclear how organisms that have evolved mechanisms adapted to living with naturally occurring nanoparticles will interact with engineered nanoparticles.

In their solid phase many anthropogenic nanomaterials look like colored talcum powders; these fine powders may under some circumstances be free in the gas phase, but they are rarely designed that way intentionally. Little to no data exists indicating whether or not nanomaterials can become aerosols. Normally, they disperse in liquids, not in the air. The mobility of nanoparticles will be influenced greatly by their tendency to aggregate or adhere to surfaces. This tendency may vary depending on the environment surrounding the particle. For example, gold nanocrystals of ten-nanometer diameter can be dispersed in water to provide truly separate nanoparticles. Small changes in the concentration of salts in the solution, however, can cause rapid aggregation of the particles into larger masses. Some nanostructures, such as carbon nanotubes, exhibit enormous attractive forces between particles, which cause them to form aggregates of much larger dimensions than the individual nanoparticles. Carbon nanotubes form compact ropes that nonetheless can be separated into individual tubes with appropriate treatment and stabilized in water and organic solvents. Intuitively, one would anticipate that aggregates of nanoparticles should be less mobile in air or in water than individual nanoparticles. The theory for aggregate transport in porous media, however, suggests that it is conceivable that aggregation into a critical size range might actually enhance particle mobility in some circumstances. Aggregation is also likely to result in an actual size of nanostructures in biological environments that could be quite distinct from the initial size of the nanoparticles.

Observations reported by others that nanoparticles can be taken up by cells suggest that accumulation in cells may contribute to the fate of nanoparticles in the environment. Possible impacts of nanomaterial uptake include:

- direct toxicity to cells
- alteration of protein conformation
- structural interference in cell division
- persistence within the cell

- the ability to transport other associated materials into cells such as contaminants or scavenged genetic material

Understanding the extent to which uptake and persistence will occur in microorganisms may provide critical insight into one important exposure pathway to higher organisms known as bioaccumulation.[7] Because nanoparticles often have high reactivity relative to their mass, they also present the possibility of mediating reactions that may influence cells. For example, carbon C60 "buckyballs" are known to be efficient photosensitizers. Under certain conditions these nanoparticles may generate oxidants that are lethal to cells. Such an effect has been considered with important medical applications such as the treatment of cancerous tumors.

Nanoparticles and colloids may facilitate the transport of otherwise immobile contaminants in groundwater aquifers and surface waters, as well as reduce the removal of particle-associated contaminants by water-treatment facilities. For example, colloid-mediated transport of plutonium waste in Nevada led to rates of radionuclide transport through groundwater orders of magnitude larger than predicted. The role of naturally occurring and anthropogenic nanoparticles in facilitating the transport of other materials is not understood.

Several unique features of nanoparticles suggest that the transport and fate of nanoparticles may differ from those of the more extensively studied colloidal systems, both quantitatively and qualitatively. One difference is size. The nanoparticles' smaller size results in a greater importance for diffusive transport and greater potential for biouptake,[8] and a unique surface chemistry. A second difference is structure. Engineered nanoparticles often display faceted surfaces that may affect their transport and reactivity.

Toxicity Research

While the field of nanotechnology is growing exponentially, a striking feature is the near absence of papers concerning the toxicology of these new materials. Casual observations from laboratories where scientists work with nanomaterials have not, thus far, suggested serious toxicity issues following nanomaterial exposure. Cells grown in culture can take up semiconductor nanoparticles without apparent ill effects. Such anecdotal evidence, however, is not conclusive toxicology. If nanomaterials behave anything like other foreign particles with respect to human health, their effects are likely cumulative and may not be apparent in individuals for decades. Only comprehensive and extensive toxicological research can answer the question: Are nanomaterials safe?

One difficulty of toxicity research on nanomaterials is that no directly relevant background studies exist. For example, *the only paper* published as of August 2002 specifically addressing the toxicity of a product of nanochemistry was a University of Warsaw study of carbon nanotube inhalation that showed no respiratory distress in guinea pigs.[9] While there is a glaring lack of data on health effects of nanomaterials related to occupational or environmental exposure, there have been a number of studies that consider the medical uses of nanomaterials and their associated toxicity. For example, progress has been made in understanding mechanisms of DNA cleavage and tumor cell death that might be exploited in cancer treatments using fullerene-based nanomaterials. Currently, these observations can only be evaluated in the context of experience and experiments that are arguably inappropriate. Humans have contended with the presence of particulate matter in their air and water for centuries. Whether the experience with naturally occurring fine particles can be extrapolated to engineered nanomaterials is currently unknown.

Nanomaterials may differ from other particulate materials in both size and surface chemistry. In brief, nanomaterials are all surface. A gram of single-walled carbon nanotubes, for example, has over ten square meters of available surface. Control over the chemistry of this interface is essential in developing nanotechnologies. Rarely will bare inorganic solids be present in the solution phase of nanostructures; rather, nanoparticles will likely be coated chemically to give them specific desired properties. Biological functionality may be integrated into a surface to draw nanoparticles into certain cells, or to disperse particles in particular organs. In natural environments, anthropogenic nanoparticles may be coated with materials such as the decay products from leaves and other biotic materials. One might speculate that the toxicology of a metal nanoparticle may be influenced more by its surface chemistry than its interior composition. This distinction is far from academic, however, as regulators must grapple with the question of whether or not conventional materials for which we have substantial toxicological data will have the same toxicological properties when they are formulated as nanomaterials.

Thus, early studies on the toxicity of nanomaterials such as the Polish study will raise as many questions as answers. Studies and variations on studies need to be repeated before definitive conclusions regarding toxicity can be drawn.

Researching the toxicology of nanomaterials will also be tedious. Research in this area will need to address the health effects for each major class of materials separately. Pure powders of isolated particles or fibers, which are the outcome of most manufacturing processes, will pose a different challenge than solid composites which have nanoscale components integrated into them. This diversity is amplified by the fact that nearly all major classes of materials can be made in a nanoscopic format: graphitic carbon-based nanostructures make up fullerenes and single-walled carbon nanotubes, metals such as gold and nickel are used to make nanoparticles and nanoshells, and ceramics and semiconductors are formed as nanoparticles, wires, and tubes. This diversity makes it essential that early toxicological research in nanomaterials focuses on deducing systematic trends from well-controlled and thorough comparisons of a selected few key material classes.

Implementing Nanotechnology as a Tool of Sustainability

Although there are many unknowns surrounding the fate of nanomaterials in the environment and their impacts on human health and ecosystems, there is very little uncertainty about the negative impacts associated with some of the materials currently used to produce nanomaterials, such as chlorinated solvents or toluene. Therefore, the growing nanomaterials industry must take pains to avoid problems of the nature that were sometimes encountered during the early growth of the semiconductor industry such as groundwater contamination by solvents used in the manufacturing process.

The potential for making nanomaterial production a "green" activity is great when such problems are addressed early in the development of this industry. Consideration should be given to minimizing energy consumption in the process, optimizing materials usage, and identifying opportunities for closing the loop on energy and materials in an industrial ecology context.

Perhaps most importantly, the education of our nanoscientists and engineers, as well as members of the larger environmental community, will likely be our primary assurance that nanotechnologies emerge as tools of sustainability. Not all risks can be anticipated. New materials will be created. New applications of these materials will be devised. The practice of an environmentally responsible industry and the implementation of nanotechnologies as sustainable technologies should not be left to environmental engineers, scientists, and economists alone. By integrating the environmental perspective into the educational programs of our chemists, chemical engineers, managers, and others, the professional cadre in this emerging industry will be prepared to better anticipate and respond to new issues as they arrive in the future. Environmental scientists and engineers must also update curricula to consider the special challenges posed by this and other emerging industries. Nanochemistry and the technologies it inspires are far too important to be left in the hands of nanochemists alone.

For Further Exploration

Environmental and Energy Systems Institute web site: www.ruf.rice.edu/~eesi/.

Nanotechnology and the Environment web site: www.environmentalfutures.org/nanotech.htm.

Notes

1. Nanotechnology is the science of manipulating matter at the nanoscale. Nano means a billionth of a meter. The basic premise of nanotechnology is that things can be built from the bottom up, created out of the building blocks of matter—atoms, as opposed to the traditional way of building from the top down. Nanotechnology combines many different disciplines such as physics, chemistry, and biology.
2. A *membrane* is a semipermeable barrier that selectively separates constituents in a

liquid or gas. Membranes can selectively separate components over a wide range of particle sizes and molecular weights and are used in various commercial applications.

3. A *catalyst* is a substance that increases the speed of a chemical reaction without itself being consumed in the process. Catalysis occurs when such a substance accelerates a chemical reaction.
4. A *spectrophotometer* is an instrument that uses different wavelengths of light to measure the concentrations and purity in a solution. For many spectrophotometers, the wavelength accuracy is plus or minus one nanometer, and the wavelength range is 190 to 1,000 nm.
5. Thin film solar cells are ten to a hundred times thinner and potentially lighter than today's silicon photovoltaic cells. Because they require less semiconductor material than other solar cells, thin film solar cells can be made for less money.
6. *Colloids* are small particles suspended and dispersed through a different medium, for example, fat droplets in milk. Colloid science deals with systems in which one or more phases, such as the fat droplets, are dispersed in a continuous phase of different compositions or states, such as milk.
7. *Bioaccumulation* is the increase in a concentration of a chemical in a biological organism over time, compared to the chemical's concentration in the environment.
8. *Biouptake* refers to the entrance of a chemical into an organism, for example by breathing or absorbing it through the skin, without regard to its subsequent storage, metabolism, and excretion by the organism.
9. A. Huczko et al., "Physiological Testing of Carbon Nanotubes: Are They Asbestos Like?" *Fullerene Science and Technology* 9 (2, 2001):251–254.

5

Ecological Computing

Feng Zhao and John Seely Brown

Great Duck Island, Maine

On a small patch of land ten miles off the coast of Maine, a team of computer engineers from the University of California—Berkeley is conducting an experiment in ecological computing. Working with biologists at the College of the Atlantic, the engineers have installed 190 wireless sensors that are being used to monitor the habitat of nesting Leach's storm petrels on the island. In the past, the biologists studying nesting behaviors of the birds had to travel to the island every now and then to gather observation data. To check on the petrels, they literally had to stick their hands into the burrows, often causing the birds to abandon their homes.

Now, these same biologists are checking on the birds on the island in the comfort of their offices, browsing data from the sensors linked by a satellite. And their colleagues thousands of miles away can share the same information thanks to the Internet. The untethered, matchbox-sized sensors left in the burrows monitor the occupancy by recording temperature variations inside and wirelessly send the data to a gateway node on the island. Convenience aside, the more significant benefit of the technology is the minimization of disturbance to the very habitat that it tries to help preserve.

The experiment on Great Duck Island is a small lens into an expansive future. To grasp what might happen, multiply these 190 sensors by 10 million or 100 million and distribute them globally. When the sensor grid becomes ubiquitous, it becomes like an enormous digital retina stretched over the surface of the planet. This planet-scale system could help us understand and address tomorrow's environmental challenges, ranging from monitoring global biodiversity to sensing millions of low-level, nonpoint sources of pollution.

Let's add intelligent browsers to this vast sensing system that lets scientists, government regulators, or environmental advocates use the Internet to ask questions never before imaginable. Call this Google on steroids. We could search vast amounts of data for abnormal events or detect interesting patterns at many different scales. But what else?

Equipped with a new generation of sensors, automobiles and trucks could monitor their own emissions and download them at a service station or to a home computer, or transmit the data in batches over cellular networks. When cars can talk to each other we can begin to create dynamic networks that can be optimized to reduce congestion, cut air pollution, speed up just-in-time deliveries, or help people find the closest available parking space in an unfamiliar city. This is more than just about convenience. We waste enough energy sitting in traffic jams each year to run our entire domestic airline fleet.

As networked sensors become dramatically less expensive and have wireless capability built into them, we may find them in a Midwest cornfield, helping farmers optimize water and fertilizer use and minimize the use of harmful pesticides. Sensor systems can go where we cannot, monitoring environmental damage in an oil spill or forest fire, tracking ocean currents, or helping biologists unravel the wonders of the rainforest canopy. We could begin to instrument whole ecosystems, using ground-based sensors networked to the next generation of satellites to understand subtle but far-reaching changes in land use and vegetation.

With pervasive and embedded intelligence, our manufacturing systems could become self-managing and more self-regulating. Products and parts with on-board sensors and radio frequency tags could keep track of themselves, help manage inventories, know when they need repair or replacement, and find their way back to the right place to be remanufactured or recycled. These systems would be capable of acting independently in response to their environment without requiring constant, and often expensive, human intervention. Industrial systems would begin to operate more like ecological ones, continually aware of their surroundings, self-organizing, and perhaps even engaging in micro auctions for balancing energy loads.

Can any of this happen? Yes. And here is why and how.

Instrumenting the Planet

In ten to twenty years, the Internet will change many of the ways that biologists and ecologists study living systems at nearly every scale. Visionaries, entrepreneurs, and techies are designing an omnipresent, planet-scale sensor network that will dwarf the Internet by many orders of magnitude. This sensor network, or informational grid, will provide entirely new kinds of instruments for doing environmental sciences on a scale never before possible. This grid will be adaptive and be able to select and attend to interesting things happening in the environment. To build a system of this size, computer scientists and engineers will have to borrow ideas from biology and ecology, and figure out how large-scale complex systems adapt, repair, and self-organize. An open, two-way interaction between environmental scientists and computer scientists is likely to have far-reaching implications for both the computational and biological worlds for many decades to come.

The transformational force underlying this change is the confluence of recent rapid technological advances such as micro-electro-mechanical system (MEMS) sensors and actuators, wireless and mobile networking, and low-power embedded microprocessors. Moore's Law describes our ability to progressively manufacture smaller and smaller transistors but that, in turn, suggests that we can progressively make sensors cheaper, smaller, more versatile, and less power hungry. Wireless sensors that integrate communication, computation, memory, sensing, and onboard power in a single package can be used to detect anything from temperature, humidity, light, sound, pollutants, and so forth, to traffic on roads. Today, one can readily order buckets of sensor nodes from startup companies such as Crossbow that manufacture integrated sensors. These matchbox-sized sensors, costing one to two hundred dollars each in small quantities today, will soon be smaller than a thumbnail and cost no more than a few cents each, when produced in massive quantities. Massively distributed systems built on these emerging sensor technologies will have to be designed to operate in radically different ways. In fact, our normal notions of personal computing will not help us much in understanding an emerging distributed system that will largely run itself.

Self-Organizing Systems

Unlike the Internet that requires round-the-clock human supervision and maintenance, the planet-scale sensor net will be for the most part autonomous, self-configuring, and attentive to its context and to its users. Because it is deeply embedded into the physical world, the system is subject to some very severe constraints, the most important of which is limited onboard energy supply. Remember, the sensors will be using wireless to talk to each other. The same principle applies: the more you talk on your cell phone, the faster you drain your battery.

To maximize the usefulness and lifetime of the system, the sensor net has to adapt (and reorganize itself) as environmental conditions or user needs change. Engineers are busy figuring out how such systems can borrow ideas such as diffusion and reaction from living organisms. Nature does it very well. For example, termites in Australia's Kakadu National Park build impressive mounds with little global knowledge or design.

Researchers have been developing biology-inspired, lightweight, peer-to-peer communication protocols link the sensors together. Unlike the Transmission Control Protocol/Internet Protocol (TCP/IP) for the Internet that uses fixed addresses to route information, the so-called diffusion routing protocols from the University of California at Los Angeles/Information Sciences Institute (UCLA/ISI) use the data that sensors collect to dynamically set up routing pathways, thus incurring much less overhead than TCP/IP. More recently, scientists at Xerox's Palo Alto Research Center (PARC) have developed a new protocol called

Information Driven Sensor Querying (IDSQ) that uses an information gradient derived from local sensor data to self-organize the network into coherent aggregates of nodes. As the physical phenomena in the environment move or application requirements change, the aggregates adapt accordingly, all without centralized supervision. A new generation of featherweight sensing, communication, and security protocols is being developed to make practical deployment of sensor nets a reality.

An important feature for such a large-scale organic system is the ability to focus on interesting things. To alleviate the problems with limited network bandwidth and limited human cognitive bandwidth, the sensor net must allocate limited resources to attend to current and emerging events of interest, while ignoring irrelevant stimuli. The mechanisms of biological vision systems can inspire interesting design here. They can reduce the amount of energy the system needs to use, since the nodes that are not in the attentional foci can be turned off until needed.

Such a system should also be able to model and calibrate itself. What does a node know about the world when it wakes up in the wild? One piece of knowledge crucial for many sensing tasks is the location of the node. Can nodes figure out their relative distances and orientations by looking at some common physical phenomena? Human vision systems can get subpixel resolution exploiting the randomness of the retina cell placement. We may be able to turn this around, and use moving stimuli in the form of scanning lines or shadows for sensors to calibrate their positions with respect to common reference stimuli.

A big concern often raised about such an omnipresent system is personal privacy. How could one walk down a street without being videoed and tracked by an unauthorized individual? One way to protect our privacy is to build some friction or loss into the system, so that not every piece of information is immediately accessible or recoverable. Ideas such as statistical sampling used in databases could be extended to allow resolution-controlled access to the various information repositories on the sensor net. Just like in a human vision system, one can imagine building some sort of defocusing lenses that blur out identity information, say license plate numbers, while providing aggregate data like the extent or size of the traffic jam on Highway 101. But, of course, how such a distributed system can be controlled and by whom are not simply answered by listing technical capabilities, especially given the self-organizing nature of this kind of system.

Ecological Computing: The Co-Evolution of the Digital and Biological Worlds

The collective challenge facing the computer science and environmental communities now is how to move from our traditional focus on personal computing to broader ecological computing that utilizes the notions of complex adaptive and self-organizing systems in the design of a new kind of information fabric. Ecological computing systems are blended into the physical environment through sensors, actuators, and logical elements; they are invisible, untethered, adaptive, and self-organizing. This is where the computational world meets the physical world.

A sensor net is an example of an ecological computing system. To coexist and coevolve with the surrounding environment, the sensor net must be able to regenerate itself, and recycle its parts for new uses. Living systems are incredibly good at this. The sensor net has to be designed in a similar way if we expect it to survive. What will you do with a dead sensor node, depleted of battery power? Changing the batteries every couple of weeks on zillions of such embedded sensor nodes, some of which may even be physically inaccessible, is clearly not feasible. Leaving the dead nodes out there will create the next environmental superdisaster that will cost our grandchildren dearly to clean up.

However, there are abundant energy sources in the environment. Sensors could harvest energy from vibrations of passing foot traffic, temperature differentials of body heat, or chemical reactions in the soil. Solar energy is a clean and limitless source. The current-generation solar panels are still too inefficient for a massive deployment for sensor nets. One may find their use in an open environment such as a desert. But forest and densely covered areas may require other modes of energy harvesting, unless one can set up the solar panel high up above tree canopies or building tops.

One interesting idea suggested by Gerry Sussman of the Massachusetts Institute of Technology is to hijack garbage-eating bacteria commonly found in woods to convert carbohydrates in the environment into sugars that can be used to power up sensor nodes. Out in the wild, there are plenty of dead tree leaves, branches, roots on the ground that are good sources of carbohydrates. If we are careful in adjusting the density of the sensor nodes and their duty cycles, nature can easily replenish what is consumed by the bacteria. An extra bonus of using such bacteria-powered energy cells is that their naturally occurring "rotten smell" can be exploited to repel certain unwelcome animals from devouring the sensor nodes without causing environmental damages. Interestingly, one of us had the misfortune of quite unintentionally attracting animals to sensors in one of our recent field experiments in the Mojave Desert. Coyotes chewed up all the windshield foam on our microphone sensors apparently because the foam contained a chemical called uria that the coyotes found tasty.

On a more global scale, balancing energy across nodes could shift the load from low-energy-reserve nodes to high-powered ones. Remember, the fabric as a whole is the sensor. There will be some nodes somewhere that have some energy left. The trick is to get the energy to where the action is. Some sort of peer-to-peer diffusion, with a gradient set up autonomously by the local energy demand, may work here. The net may even be able to heal itself, patching sensing holes or moving nodes around. Nanotechnology might enable damaged nodes to grow new "eyes, ears, and noses." Clearly, environmentally friendly design must be high on any future agenda for such systems.

Programming an ecological computing system will be more like designing a biology experiment, telling the system to "grow a finger here," than writing low-level embedded programs. New transformations need to be invented that translate global properties one wants to design into some local representations that are easy to specify and implement on the embedded nodes. Invisible "compilers" will take care of the low-level, mechanical translations. Computer scientists and environmental researchers must join hands in the design of the programming technology.

Various species in nature coexist through some very complex dependencies and feedback loops, the so-called web of life. The coming digital and ecological worlds will coevolve in a similar, symbiotic way. In fact, the boundary between the two will be so blurred that we may not even be able to tell one from the other in this ecological computing fabric. Of course, at this juncture we are not certain how far this symbiosis can go but we are certain that as we move from just focusing on information processing systems to massively distributed informating systems—systems that read and respond to their context—we will have fundamentally new tools for analyzing and effecting ecological systems. How such systems are designed and applied to the environmental challenges of the future is one of the primary governance challenges of today.

For Further Exploration

Martha Baer, The Ultimate on-the-Fly Network. *WIRED*, December 2003.

J. S. Brown and D. Rejeski, Ecological Computing. *The Industry Standard*, December 18, 2000. At: www.thestandard.com/article/0,1902,21365,00.html?printer_friendly=.

Crossbow Technology, Inc. web site: www.xbow.com.

Gregory Huang, Casting the Wireless Sensor Network. *Technology Review*, August 2003.

National Research Council, *Embedded, Everywhere: A Research Agenda for Networked Systems of Embedded Computers*. Washington, D.C.: National Academy Press, 2001. At: www.nap.edu/html/embedded_everywhere/.

RAND corporation web site (see the chapter on "Beyond the Internet"): www.rand.org/scitech/stpi/ourfuture/.

Xerox PARC Collaborative Sensing Project web site: www.parc.com/ecca.

6

Genetics and the Future of Environmental Policy

Gary E. Marchant

Just a few years into the twenty-first century, it is already being described as the century of the gene. Advances in genetics will have major medical, economic, ethical, political, and personal implications that will affect every segment and aspect of society. Environmental policy will not be exempt from this revolution, and will almost certainly be radically transformed by new genetic technologies.

Consider some of the genetic advances predicted for the next couple decades. Dr. Francis Collins, director of the Human Genome Project, predicts that testing for genetic predispositions will become standard practice, and individualized preventive medicine will be available based on those genetic susceptibilities. Dr. French Anderson, a leading expert on gene therapy, predicts that by 2030 gene therapy will be available for essentially every disease. Moreover, he predicts that gene therapy will also be used to *prevent* future disease by altering individual genetic weaknesses affecting our susceptibility to disease.

Many of these genetic advances will have direct applications to environmental protection and increasing relevance in a world of higher resource productivity described by many of the authors in this volume.

Four potential scenarios demonstrating how genetics might affect environmental policy are presented below.

Scenario 1: Toxicological Characterization of Environmental Substances

Only a small proportion of the approximately eighty thousand chemicals in commerce have been thoroughly tested for safety. The current gold standard for toxicity testing is the chronic rodent bioassay, which takes three to four years to complete and costs upwards of $3 million.[1] Chronic cancer bioassays have been completed for only a small percentage of the chemicals in commerce, and as a practical matter most chemicals will never be tested given the enormous costs, resources, and time required for such tests. Even fewer chemicals have been subjected to the full battery of other toxicity tests necessary to fully evaluate safety, including tests for genotoxicity,[2] teratogenicity,[3] reproductive effects, neurotoxicity,[4] systemic toxicity, and endocrine disruption. In short, we are doomed to operate in toxic ignorance with current testing technologies.

An important new genetic technology called a DNA chip or microarray will revolutionize toxicity testing.[5] Microarrays are already being marketed that contain all human genes arrayed in specific locations on two chips the size of a postage stamp. These microarrays can be used to profile the expression of genes in a cell after exposure to a toxic chemical. Every toxic response appears to be associated with a specific pattern of changes in gene expression, providing a molecular fingerprint of exposure to that chemical and its toxicological mechanism. Initial studies have demonstrated that microarrays can successfully characterize the toxicological potential and mechanism of chemicals based on their gene expression profiles.

Within a decade, microarrays will be the standard technology for evaluating toxicity. It only takes a few hundred dollars and a couple days to test a chemical using microarrays, a considerable advance in both speed and cost over chronic bioassays. Microarrays offer other important benefits compared to current test methods. All toxicological effects can be evaluated in a single microarray assay, whereas today separate tests are needed to evaluate each potential effect (e.g., cancer). Microarrays are also more sensitive than current methods because they can detect immediate changes within every exposed cell or organism, whereas existing methods can only detect observable toxicity that develops many weeks, months, or even years after exposure, and in only some cells or organisms.

To be sure, there remain important uncertainties and validation steps before microarrays can replace conventional toxicity tests. Nevertheless, the development and use of microarrays is advancing at an unprecedented pace in toxicology. Using this technology, it will soon be possible to quickly and cheaply obtain comprehensive and accurate toxicity data on every substance in commerce, a dramatic departure from today's toxic ignorance. Based on this data, it will be feasible to eliminate substances that present unacceptable risks, including substances whose toxicity would not otherwise be detected until after they had imposed significant impacts on public health. Product manufacturers will also be able to use these tests to screen out risky compounds in the product development cycle before they have invested significant resources (a process that is already occurring in the development of pharmaceuticals).

Yet, in some ways microarrays might produce more information than we can comfortably handle. Given the tens of thousands of chemicals in our environment, it will not be feasible or even desirable to try to eliminate all the substances with some toxic potential. Many will present only *de minimis* risks, others will be indispensable to modern life, and still others will be impossible to eliminate because, for example, they occur naturally.

How will citizens respond to information that they are routinely exposed to carcinogens and other potentially toxic substances? Public opinion surveys and risk perception studies consistently find a strong public sentiment for eliminating rather than managing carcinogenic and other toxic substances. To date, society has been unable to come to consensus on any measure of acceptable risk. The availability of comprehensive data on the toxicity of virtually every substance in the environment is likely to spawn bitter debate and controversy, not as today about whether particular chemicals are or are not hazardous, but rather about which known risks we should tolerate and which risks we should target for elimination. The technological advance offered by microarrays for characterizing toxicity will therefore provide an unprecedented amount of information on risks. This information will benefit public health, but will also present difficult legal, social, and ethical challenges in deciding what risks we are willing to tolerate, both as individuals and as a society.

Scenario 2: Genomics in the Courts

DNA microarrays will also be used to monitor risks to specific individuals and populations. Workers in hazardous workplaces, residents living near a polluting facility, and consumers using a potentially risky product can be monitored for changes in their gene expression. The use of microarray scans may eventually become a standard component of annual medical checkups. A physician will take a blood sample, analyze it using microarrays for any abnormal gene expression patterns, and alert the patient of any potentially toxic exposures of which the patient may not even be aware. Early detection of toxic exposures can prompt individual and public health interventions, ranging from removing the individual from the hazardous environment to providing treatment of the early stages of disease, conducting ongoing medical monitoring, or instituting regulatory or mitigation programs to reduce risks. Some of these interventions, such as removing individuals from nearby exposure threats, would constitute significant changes in the way we presently view the role of environmental protection and law.

The use of microarrays to identify individuals at risk from environmental exposures may confront courts with a major challenge. A growing number of litigants who have been exposed to toxic substances, but who have not yet manifested clinical disease, are already bringing lawsuits to recover for their latent risks. Some seek compensation for the increased risk itself, others for their fear of disease as a result of exposure, and still others for funds to monitor their health on an ongoing basis. While some of these claims have been successful, courts generally have imposed rigorous evidentiary obstacles because of concerns that a flood of asymptomatic litigants seeking recovery for their hazardous exposures would overwhelm the judicial system. Many courts have thus required plaintiffs seeking recovery for latent risks to prove that they have a present injury and/or to sufficiently quantify their risk of future disease. Most asymptomatic litigants exposed to hazardous substances cannot meet these requirements with existing technology.

Using microarrays to screen for gene expression alterations may overcome the legal barriers established by the courts and result in a deluge of latent risk claims. Some courts may treat the objective scientific evidence of changes in gene expression provided by microarrays as a "present" injury, especially those that have already ruled that subcellular genetic changes can constitute a present injury for purposes of recovering for latent risks. In addition, microarray data should permit fairly precise quantitative estimates of risks from specific changes in gene expression. Microarray data may therefore tear down the existing legal obstacles to recovery for latent risks, potentially overwhelming courts and product manufacturers with such claims in the future.

Scenario 3: A Shift to Individualized Self-Help Measures

The deciphering of the human genome has revealed extensive interindividual variation in many of the genes coding for enzymes that metabolize foods, drugs, environmental agents, and other exogenous compounds that enter our bodies. For many of these genes, between 1 and 50 percent of the population carries variations (known as "polymorphisms") of the normal version of the gene. Each of us therefore has a unique set of genetic polymorphisms that affects our susceptibility to environmental exposures.

These interindividual differences in susceptibility will become increasingly important for environmental policy in the future. In a society with high resource productivity and micro or zero emissions, most environmental exposures will result from residual low-level discharges that are harmless to most people or through exposure associated with the use of products. As environmental exposures are reduced, genetically susceptible subpopulations who remain hypersensitive to low-level exposures will become an increasingly large proportion of those remaining at risk.

In many cases, it may not be economically feasible to adopt more stringent societywide environmental standards to protect a relatively small percentage of the population susceptible to a particular pollutant. By the year 2020, more targeted environmental protection measures will become necessary to protect these susceptible subpopulations. For example, products will carry warnings notifying those carrying particular susceptibility genes that they are at increased risk from exposure to the product. Companies will be required to test their products for effects on susceptible subgroups and provide appropriate warnings.

This shift to individualized self-help measures based on susceptibility warnings has already begun. In 1999, a vaccine manufacturer was sued for failing to test its product on genetically susceptible individuals, for failing to warn of an alleged genetic susceptibility within the population to its vaccine, and for failing to recommend that consumers obtain a genetic test to determine their susceptibility before administering the vaccine. Some products today already contain warnings based on genetic susceptibility, such as the warnings on diet sodas for people with the genetic disease phenylketonuria.

Warnings based on genetic susceptibilities will only be effective if individuals are aware of their genetic susceptibilities. Commercial genetic test kits are already being marketed directly to consumers interested in discovering their genetic susceptibilities to foods and other potentially hazardous exposures. By 2020, testing for such genetic predispositions will be the norm. As people discover their unique genetic profiles, it may become more effective to rely on individualized knowledge and lifestyle changes rather than societywide environmental standards to protect against many exposures.

This shift from government-imposed regulation to individualized self-help measures will raise many controversial issues. For example, when is it appropriate to shift the burden for environmental protection from government regulators and product manufacturers to susceptible individuals? If the burden is placed on the individual, what happens when the susceptible individual fails to take the appropriate preventive or avoidance measures? Can insurers refuse to reimburse recalcitrant individuals for their health expenses on the theory that the individual knowingly placed himself or herself at risk? Can the individual file a tort suit against the product manufacturer, or should private liability suits be preempted when there has been a regulatory judgment that the individual rather than the manufacturer should bear the burden of avoiding exposure or harm? Once again, advances in genetic technology will offer new opportunities and approaches to environmental policy that will be accompanied by difficult legal, ethical, and social dilemmas.

Scenario 4: The Genetic Enhancement of Toxic Resilience

Perhaps a little further into the future, it may be possible to genetically alter humans to make them more resilient to environmental exposures. Many genetic susceptibilities to environmental exposures are due to deficient or inactive genes for particular enzymes. For example, approximately 50 percent of Caucasians are missing both copies of a gene (*GSTM1*) coding for one form of a metabolically important family of enzymes known as the glutathione S-transferases. This deficiency is associated with an increased risk of bladder and lung cancer from exposure to several toxic substances normally detoxified by the missing enzyme.

Individuals lacking such enzymes might be treated by adding a normal copy of the missing gene to their genome. In addition to such prophylactic interventions, it may also be possible to add new genetic material to repair otherwise normal genes damaged by past toxic exposures. Initially, such interventions would likely be in the form of somatic gene therapy, in which copies of the missing gene are inserted into nonreproductive cells of an existing person, and thus have no possibility of being passed on to future generations. This approach would have important limitations. Barring the development of a novel gene delivery system capable of reaching all tissues, somatic gene therapy will likely be restricted to cells in easily accessible tissues such as skin and blood. This would provide little benefit for exposures that exert their primary toxic effects in critical tissues such as the liver or lung. In addition, many somatic gene therapies will require successfully altering the genes of millions of individual cells to be effective.

Given these limitations, it is perhaps inevitable that germline gene therapy will be attempted. In this case, a single embryo cell could be genetically transformed with environmentally protective genes and then grown into a person in which every cell contains the new genetic information. Because a cell from an embryo created by in vitro fertilization can be isolated and treated relatively easily in a petri dish, germline gene therapy will be much easier to accomplish than trying to deliver genetic information into the tissues of an existing person.

One approach likely to be attempted to genetically enhance environmental resilience is to add an artificial chromosome into a developing embryo. Artificial human chromosomes have already been created and inserted into stable human cell lines grown in tissue culture. Artificial chromosomes are an attractive option because they do not require insertion of genetic material into existing chromosomes of the cell, which may disrupt the normal functioning of nearby genes. Instead, the artificial chromosomes establish themselves as stable, freestanding autonomous units within a cell's nucleus. It may be possible to add different sets or suites of genes clustered together on interchangeable modules that can be inserted onto the artificial chromosome.

In the future, fertilized embryos may routinely be genetically tested for their complement of environmentally important metabolism genes. An artificial chromosome containing a suite of protective genes missing or defective in the embryo could be selected and added to the embryo to enhance its genetic resilience, essentially producing an "environmentally protected" individual. Germline gene therapy and genetic enhancement, if successful, would likely not be limited to environmental resilience, but may also include genetic improvements ranging from resistance to various diseases to enhancement of intelligence, personality traits, or skills. These applications will no doubt be controversial, but public opinion polls consistently show strong support for genetic modifications that protect against future disease (but not eugenic improvements to intelligence or personality).

An expert panel convened by the American Association for the Advancement of Science issued a report opposing inherited genetic modifications (IGM) in general, but noted that the use of IGM to prevent and treat clear-cut diseases in future generations is ethically justifiable. Many parents of an embryo that would otherwise carry important genetic susceptibilities to environmental exposures will have strong incentives to take whatever measures are available to enhance their future child's resilience and health. Gregory Stock, author of *Redesigning Humans*, predicts that certain genetic enhancements will be available that any responsible parent would make, just as we generally agree that kids should enhance their immunity by getting vaccinations.

Conclusion

Some or all of these potential applications of genetics to environmental policy may seem far-fetched. There will no doubt be many complications and obstacles to overcome in each of the scenarios described above. Yet, if experts were asked to predict the advancement of genetics just ten years ago, few would have predicted that we would have already sequenced the human genome, identified and characterized hundreds of susceptibility genes, developed DNA chips that can quickly and cheaply screen for thousands of genes simultaneously, genetically engineered many foods, cloned sheep and other animals, and successfully practiced preimplantation genetic analysis. The advancement of genetic science is proceeding at an exponential pace, and while any predictions of future technology are always precarious, the scenarios outlined above illustrating the revolutionary impacts of genetics on environmental policy may well turn out to be too conservative.

For Further Exploration

Stephen H. Friend and Roland B. Stoughton, The Magic of Microarrays. *Scientific American*, February 2002:44–53.

Muin J. Khoury, Wylie Burke, and Elizabeth J. Thomson, *Genetics and Public Health in the 21st Century*. New York: Oxford University Press, 2000.

Gary Marchant, *Genomics and Environmental Regulation: Scenarios and Implications* (2002). Available under "Papers" at: www.environmentalfutures.org/genomics.htm.

Kenneth Olden and Samuel Wilson, Environmental Health and Genomics: Visions and Implications. *Nature Reviews Genetics* 1 (2000):149–153.

Gregory Stock, *Redesigning Humans: Our Inevitable Genetic Future*. New York: Houghton Mifflin Co., 2002.

Notes

1. A bioassay is a determination of the strength or biological activity of a substance, such as a drug or hormone, by comparing its effects with those of a standard preparation on a culture of living cells or a test organism. A chronic rodent bioassay would be the testing of a particular substance's toxicity on rodents for a lengthy period of time, typically two years, and usually focuses on the substance's potential to cause cancer.
2. The degree to which a chemical or other agent damages cellular DNA, resulting in mutations or cancer.
3. The ability to cause defects in a developing fetus. This is distinct from mutagenicity, which causes genetic mutations in sperms, eggs, or other cells.
4. The degree to which a chemical or other agent damages or destroys nerve tissue or alters some aspect of the nervous system.
5. Also called a "gene chip," this technology allows quick and inexpensive exploration of irregularities in a gene that might signal a predisposition to a particular disease. It can also provide a reading of how a gene is responding to particular stressors or exposure to a drug or toxin.

7

The Future of Manufacturing: The Implications of Global Production for Environmental Policy and Activism

Timothy J. Sturgeon

Two distinct visions compete for the future of manufacturing. The first is a vision of distributed manufacturing, in which highly automated, small-scale manufacturing plants crank out customized products where and when they are needed. Since products would be customized to meet the exact needs of the people who order them, typical lot sizes could be as small as one. The alternate vision is one of global manufacturing, in which a handful of giant plants serve world markets.

These visions share at least two assumptions. First, both assume a split between those firms that design and market products and those firms that actually do the manufacturing. Vertical integration is consigned to the dustbin of history and with it, as I will argue, traditional approaches to environmental protection tied to vertical organization. Second, both models are based on the idea that highly complex information, especially design information, can be codified and seamlessly exchanged between firms according to widely agreed-upon standard protocols. When a firm develops a new product, design information, including data relevant for environmental compliance, can be handed off into a supply base that immediately knows what to do with it. Clearly, information technology—for things like product design, supply-chain management, inventory control, and factory automation—plays a large role in both visions of the future.

The real difference, then, has to do with scale and location. Will advanced automation and better codification schemes mean that manufacturing plants will become smaller and be distributed according to the location of demand? Or will these same advances allow manufacturing to aggregate into ever larger plants located in places like China with the very lowest operating costs? These two competing scenarios have significant environmental implications. In this essay I argue that scale and cost still matter. Recent developments in industries like electronics and autos show that the trajectory of manufacturing is rapidly shifting toward the global model. The concept of distributed manufacturing, while appealing in many ways, has begun to take on the air of a 1960s-era *Popular Science* article rather than a viable vision of the future. The day when each of us has a flexible manufacturing plant in our basement or in our neighborhood seems far away indeed. The day when nearly everything we buy is made in China by a handful of huge contract manufacturers seems to be almost upon us, and because of the immediacy of this possibility, it needs to be examined from an environmental standpoint.

If indeed this global model is winning, what are the implications for the environment and for environmental activism and policy? At first blush global production might appear a lamentable trend. Manufacturing is moving to locations with lax environmental standards, out of reach of our national policy and activist communities. How responsive will huge global factories be to our calls for a healthy planet? Would it not be better for the environmental health of the planet if production remained close to home where we could keep a close eye on it? Would not small community-based manufacturers be more interested in environmentally sustainable production methods? I argue the opposite. It is my contention that global manufacturing as it is currently unfolding has some real advantages for the goal of making manufacturing more environmentally friendly, provided the implications of this restructuring are recognized early and clearly understood by the environmental community. Before I make this case, however, I will lay out the features of the emergent system of global production in some detail.

The Concept of Value Chain Modularity and the Rise of the Global Supplier

Innovations in industrial organization have always been important to economic development. Although the greatest academic and popular focus has been on how technological innovations—the steam engine, the railroad, the telephone, the computer—destroyed old industries, created new industries of their own, and enabled new developments in other industries, new systems of industrial organization have been behind many of the most important advances in economic history, from the English factory system to Henry Ford's mass production or Toyota's lean production system. A notable feature of both technological and organizational innovations is that they arise in particular locations before being adopted, and often transformed, in other places. I argue that a new system of industrial organization, the *modular production network*, has recently come into view, driven largely by the strategies of firms based in North America. As this new model spreads, so will its environmental effects and implications. This change in industry structure has been entwined with the spatial processes of geographic dispersion, relocation, and regionalization—in short, globalization.

Value chain modularity emerged during the late 1980s and 1990s from the breakup of vertically integrated corporate structures and the horizontal aggregation of activities around specific sets of closely related value chain functions. Out of this change two broad sets of firms can be identified, *lead firms*, focused on product development, marketing, and distribution, and sometimes some late-stage manufacturing steps such as final assembly, and *turn-key suppliers* focused on selling, as services, many of the value chain activities that lead firms have decided to outsource. Figure 7.1 presents a simple conceptual map of the shift from the vertically integrated organizational form of the modern corporation to the value chain modularity that characterizes the modular production network of the twenty-first century. Note that research and development, including environmental research and development, remains a vital function for each firm in the modular network, where it is specialized into product and process applications.

In electronics, for example, firms as different as Hewlett-Packard and

Ericsson have sold off most of their worldwide manufacturing infrastructure to turn-key contract manufacturers such as Solectron, Flextronics, and Celestica. In semiconductors we see the emergence of "fabless" design firms that rely on semiconductor "foundries" located in faraway places such as Taiwan. In the auto industry, Ford and General Motors have retained vehicle design and final assembly, but spun off their internal components divisions making things such as dashboards and seats into the independent suppliers Visteon and Delphi. They have also outsourced an increasing volume of component and module design and production to first tier suppliers such as Lear, Johnson Controls, Magna, and TRW.

I call these new arrangements *value chain modularity* because distinct breaks in the value chain tend to form at points where information regarding product and process specifications can be highly formalized. Linkages based on codified knowledge provide many of the benefits of arms-length market linkages—speed, flexibility, and access to low-cost inputs—while allowing for a rich flow of information be-

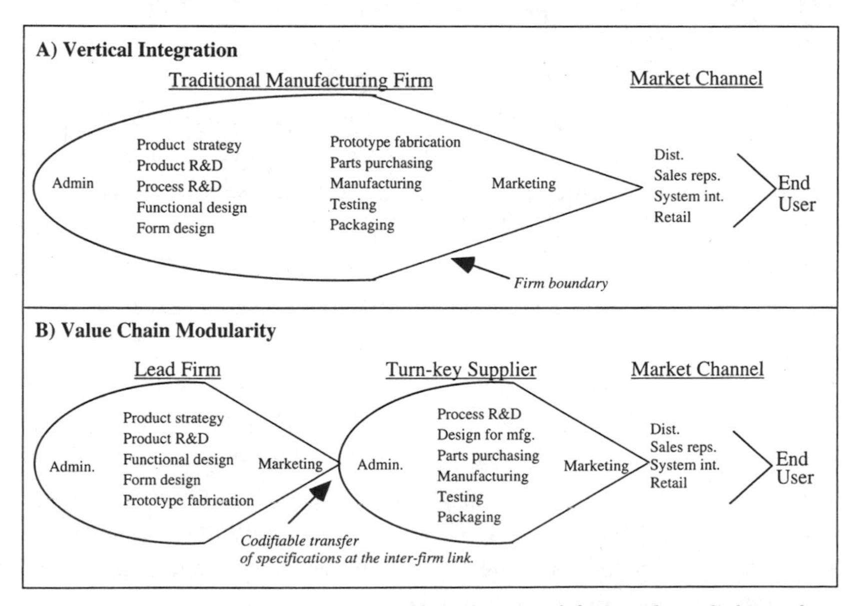

Figure 7.1 From Vertical Integration to Value Chain Modularity: The Delinking of Product Innovation from Manufacturing in the Modular Network

tween firms, including information affecting the environmental performance of modules and associated products and processes. Such transactions, however, are not the same as classic market exchanges based on price. When a computer-aided design file is transferred from a lead firm to a supplier, for example, much more flows across the interfirm link than information about prices. The locus of these value chain break points appears to be largely determined by technical factors, especially the open and de facto standards that determine the protocol for the handoff of codified specifications. The network architecture that arises from such linkages has many of the advantages of modularity in the realm of product design, especially the conservation of human effort through the reuse of system elements—or modules—as new products are brought on-stream.

The turn-key suppliers in this model contain generic productive capacity that can easily be redeployed to serve a range of lead firms as conditions change. The term *turn-key* stems from the large size, broad capabilities, and independent stance of the largest suppliers. The services of turn-key suppliers are generally available to lead firms on short notice with relatively little prior interaction. The fluidity of the network is supported by the ability to hand off relatively codified product and process specifications at the interfirm link, which has the effect of reducing asset specificity and making suppliers and lead firms substitutable. Because the production capacity that resides in turn-key suppliers is relatively generic, it is essentially shared by the supplier's customers. Environmental compliance, research, and monitoring are becoming just another arrow in the supplier's quiver of turn-key services. This is a significant departure from traditional models where environmental management functions have resided in the lead firms who have then driven environmental compliance through their supply chain using tech transfer schemes, or environmental management systems like ISO 14000.

The process of industry reorganization that has given rise to modular production networks has been occurring at the same time that firms from nearly all advanced economies, and many developing economies, have been increasing their direct and indirect involvement in the global economy. International production has long been a hallmark of American firms. Today, because of *deverticalization*, global reach is increasingly achieved with the help of a wide range of intermediaries, partners, and suppliers who support and even proxy for lead firms in far-flung locations. To try to solve some of the coordination problems that have arisen from outsourcing to multiple partners in a growing number of locations, many American lead firms have active programs in place to outsource to fewer, larger suppliers. In complex assembly industries, such as electronics and autos, this confluence of deverticalization, outsourcing, and supply-base consolidation has created a new set of important actors in the global economy, the "global suppliers."

Global Suppliers: A Key Point of Environmental Leverage in the Future?

Global suppliers introduce a high degree of value chain modularity into industry structure because the large scale and scope of their operations create comprehensive bundles, or modules, of value chain activities that can be accessed by a wide variety of lead firms. To put it differently, global suppliers are nearly always turn-key suppliers and one bundle of value-added services of interest to lead firms in the future will be environmental compliance or beyond compliance characteristics of modules and components that these suppliers produce. Value chain modularity can be conceptualized and observed entirely at the local level, but in practice industry reorganization has become deeply entwined with the processes of globalization, and it is global suppliers who perhaps best exemplify this connection.

The combination of productive scale, geographic scale, and value chain scope that has been achieved by global suppliers in only a few short years is striking. Solectron, a provider of manufacturing and engineering services for the electronics industry, grew from a single Silicon Valley location with 3,500 employees and $256 million in revenues in 1988 to a global powerhouse with more than 80,000 employees in fifty locations and nearly $18 billion in revenues in 2000. During the same period Solectron increased its offering of services related to its traditional manufacturing function to include, among others, product (re)design-for-manufacturability, component purchasing and inventory management, test routine development, final assembly, global logistics, distribution, and after-sales service and repair. Lear, a Detroit-based automotive seating supplier that generated $1.1 billion in revenues in 1991, grew to $14.5 billion in annual sales by 2000, and now operates four hundred plants in thirty-three countries and employs 125,000 workers. At the same time, Lear expanded its product offering to include entire automotive interior systems, including headliners, carpets, cockpit modules, and interior panels.

Because of the important role that American firms have played in the twin processes of deverticalization and globalization, both as lead firms and as suppliers, I characterize value chain modularity as a new American model of industrial organization. American lead firms have led the charge to outsource manufacturing and, perhaps surprisingly, the vast majority of the largest contract manufacturers and first tier suppliers that have come to dominate world production are American in origin and retain command and control functions at home. What is important about the geography of global suppliers is that they are firmly embedded in locations with both low and high operating costs. Locations in advanced economies support the set of important interactions between lead firms and suppliers that resist codification, such as codesign, prototype development, and manufacturing processes validation. Such locations are also used for the manufacture of high-value or low-volume products, or both. Engineering changes tend to occur more frequently in such products and their high unit value makes the marginal savings garnered by low-cost locations less compelling. But for an increasing array of products, production is transferred to low-cost locations—China, Mexico, or Eastern Europe—almost as soon as the manufacturing pro cess is finalized. This is possible because the plants owned by global suppliers contain identical production equipment worldwide, systems for global logistics and inventory management have improved, and transportation costs continue to fall.

The Environmental Challenges and Opportunities

What are the implications of value chain modularity and the rise of global suppliers for the environment, and for environmental policy and activism? Widespread automation and an increased focus on product quality will mean that only the most advanced production methods will be put to use, even in locations with low operating costs. Advanced production techniques, because they tend to be deployed first in places with strict environmental regulations and because they often focus on reducing waste, are typically associated with better environmental practices. Moreover, the tight integration of far-flung plants means that production methods must be nearly identical in all locations. Plants must be effectively interchangeable. This drives environmental standards toward the highest common denominator. Perhaps most importantly, consolidation of larger portions of the world's manufacturing in a handful of large contract manufacturers and huge first tier turn-key suppliers provides important points of leverage for policymakers and activists. Pressure brought to bear on a global supplier is pressure brought to bear on a substantial chunk of the world's manufacturing base.

This argument may sound familiar to activists who have had some success in upgrading working conditions in some shoe and apparel factories by pressuring Nike and to those who have helped to improve the lot of the world chicken population by pressuring McDonald's. However, these may be only one of many new environmental leverage points with ever expanding global reach. The fact that the most important global suppliers are American in origin, at least in important manufacturing industries such as electronics and autos, means that pressure applied by American policymakers and activists can be highly effective. Pressure applied at home will have an increasingly global reach. If this view sounds too glib, think instead of how hard it would be to apply pressure on a highly fragmented supply base. Since plants would not be tightly integrated and interchangeable, production methods could vary widely by location. Monitoring would be extremely difficult, and if a problem were to be identified and corrected, it would impact only a tiny fragment of world production. Global production certainly has its risks and pitfalls, not least for the firms involved and especially for manufacturing workers in advanced economies, but speeding environmental degradation may not be among them. This is especially true if the environmental community can recognize this structural transformation of manufacturing early and take advantage of the new opportunities it provides for environmental protection.

For Further Exploration:

A. Amin and N. Thrift, Neo-Marshallian Nodes in Global Networks. *International Journal of Urban and Regional Research* 16 (1992):571–587.

E. Bonancich et al., *Global Production: The Apparel Industry in the Pacific Rim.* Philadelphia: Temple University Press, 1994.

W. Davidow, *The Virtual Corporation: Structuring and Revitalizing the Corporation for the 21st Century.* New York: Harper Collins, 1992.

P. Dicken, *Global Shift: Transforming the World Economy*, 3rd ed. New York: Guilford Press, 1998.

Gary Gereffi, The Organization of Buyer-Driven Global Commodity Chains: How U.S. Retailers Shape Overseas Production Networks. In G. Gereffi and M. Korzeniewicz (Eds.), *Commodity Chains and Global Capitalism.* Westport, CT: Praeger Publishers, 1994. 95–122.

Gary Gereffi, John Humphrey, and Timothy Sturgeon, The Governance of Global Value Chains: An Analytic Framework. Forthcoming in the *Review of International Political Economy.*

M. Schilling, Towards General Modular Systems Theory and Its Application to Inter-Firm Product Modularity. *Academy of Management Review* 25 (2, 2000):312–334.

T. Sturgeon, Modular Production Networks: A New American Model of Industrial Organization. *Industrial and Corporate Change* 11 (2002):3.

T. Sturgeon and R. Florida, Globalization, Deverticalization, and Employment in the Motor Vehicle Industry. In Martin Kenney and Richard Florida (Eds.), *Locating Global Advantage.* Palo Alto, CA: Stanford University Press, 2003. 52–81.

T. Sturgeon and R. Lester, The New Global Supply-Base: New Challenges for Local Suppliers in East Asia. In Shahid Yusuf, Anjum Altaf, and Kaoru Nabeshima (Eds.), *Global Production Networking and Technological Change in East Asia.* New York: Oxford University Press, 2004. Chapter 2.

D. Teece, Firm Organization, Industry Structure, and Technological Innovation. *Journal of Economic Behavior* 31 (1996):193–224.

S. Thomke and D. Reinertsen, Agile Product Development: Managing Development Flexibility in Uncertain Environments. *California Management Review* 41 (1, 1998): 8–30.

8

Engineering the Earth

Brad Allenby

Let us be clear about one point: environmental policy is increasingly irrelevant and dysfunctional. In its command and control guise it remains necessary, especially in developing countries, but hardly innovative; attempts to extend it through, for example, mutually reinforcing cults of sustainability have been less than successful to date. This does not mean that the underlying perturbations are nonexistent or do not require our attention—quite the contrary. It does mean, however, that the dead hand of past battles and ideologies dominates the present, giving as of yet little indication that a socially and environmentally preferable outcome is in the offing. Moreover, those that are today's environmental professionals and activists may be, ironically, the least capable of building tomorrow's environmental competencies. Environmentalism as an institutionalized (and somewhat elitist) counterculture to the dominant market paradigms of the present is comfortable for all concerned, but is less and less defensible intellectually.

The first challenge, as always, is to perceive as clearly and objectively as possible the context within which we must think about our future. Here, one point becomes clear: a principle result of the industrial revolution and its associated changes in human demographics, technology systems, cultures, and economic systems has been the evolution of a planet in which the dynamics of most major natural systems are increasingly dominated by human activity. All surface waters or atmospheric nodes have in some way been affected by human activity, at least in terms of chemical composition. Likewise, no biological communities of any size have been unaffected, one way or another, by the activity of a single species—ours. Genetic engineering digitizes and commoditizes the genome, and will make all of life itself a designed system. If we leave a genome alone, it is by choice and intent. Critical dynamics of most major chemical cycles—nitrogen, carbon, sulfur, and the heavy metals—are tightly integrated into human economic and technological systems. Regional systems—the Baltic Sea, the Everglades—are now designed systems. To the extent biological systems are preserved, it is by choice—and if that choice is not made, they are not preserved. Unless they are included in the design objectives and constraints implicit in the project they disappear. Forget terraforming Mars—welcome to the anthropogenic earth. And forget "natural history"—increasingly there is only human history. And that trend will only intensify: the evolution of information technology, economic structures and globalization, nanotechnology, and biotechnology will have far, far more to do with the structure of the future than any environmental policy we may think about.

Under such circumstances, for the most part not perceived by policymakers or the public, continued stability of both human and "natural" systems requires the ability to rationally and ethically design and manage coupled human-natural systems in a highly integrated fashion—an earth systems engineering and management (ESEM) capacity. Such complex systems cannot be "controlled" in the usual engineering sense; rather, ESEM is a design and management activity predicated on continued learning and dialogue with the systems of which the engineer is an integral part. Moreover, these systems all have coupled biological, physical, scientific, technological, economic, governance, and cultural dimensions and uncertainties. Accordingly, no single discourse around science, philosophy, or economics, or set of discourses, is sufficient on its own to offer comprehensive policy guidance. Moreover, designing and managing such systems raises profound ethical and religious questions. For example, implicit in the negotiations over climate change mitigation is the question, "What kind of world do you want?" This is clearly a normative, not an objective, question. And that same question is implicit in the ongoing evolution of biotechnology, the information society, nanotechnology, or the policy choice to make egalitarian development around the world possible and bring it about.

Clearly, we do not yet have the data, tools, mental models, intellectual frameworks, ethical and religious understandings, or institutional capacity to create ESEM at this point. Accordingly, ESEM is best thought of as a capability that must be developed over a period of many years, than as something that we can implement in the short term. But that does not mean that it springs from nothingness. Rather, it is a coming into consciousness of patterns of human behavior and interrelationships with natural systems that have been going on for centuries but are just beginning to be perceived, primarily because of the scale of population and economic growth during the twentieth century. Operationally, ESEM builds on practices and activities that are already being explored. From a technical perspective, these include a number of methods and practices currently lumped under the rubric of industrial ecology, including design for environment or life cycle assessment. From a managerial perspective, it draws on the literature about managing complexity and learning organizations. From a natural resources management perspective, it draws on the new study of adaptive management developed mainly by scientists and ecologists working on complicated issues such as restoration of the Everglades or the Mississippi Delta, or management of the Baltic Sea.

A few examples might illustrate some of the critical elements of ESEM. Let us briefly examine the Everglades. It is a unique ecosystem that has been altered by an 1,800-mile network of canals built over the last century to support agricultural and settlement activity. Invasive species such as the melalu, Brazilian pepper, and Australian pine are outcompeting native species, frequently aided by human activities such as draining marshes or more frequent fires. Indeed, the natural cycles that once defined the Everglades, including rainfall and water distribution patterns and nutrient cycles, have been profoundly affected not just by human settlement patterns, agriculture, tourism, industry, and transportation systems, but also by the various management regimes attempted over the past hundred years. Not surprising, then, that the nesting success of birds, a predominant animal form in the Everglades, has declined some 95 percent since the mid-1950s. In response to this obvious change in systems state, a $7.8 billion Everglades "restoration" project has begun. Its intent is to restore waterflow in the swamp to functional levels, while continuing to support industrial, agricultural, settlement, and other human activity at politically acceptable levels.

Even this simple example is fairly instructive. To begin with, it is apparent that the Everglades has been for some time, is now, and will continue for the foreseeable future to be, a product of human design and human choice. There is no pristine state to return to; the Everglades will never be natural again. It will be an engineered system, and it will display those characteristics—including preservation of birds and other flora and fauna, if that is a design objective—that humans choose. ESEM does not imply an artifactual world. It does require that humans consciously accept responsibility for their designs, and their actions, even (perhaps especially) when part of that design is the maintenance of wild or natural areas. Moreover, the Everglades, like any complex system, will continue to evolve, and its human designers and managers will have to continue to work with it in light of the constraints and objectives our society, and systems behavior, imposes on their project. Unlike most engineering projects, this one does not end, nor does the practice of ESEM generally. Absent collapse of the human or natural systems involved, the dialogue between human and natural systems will continue indefinitely. We can do it better, we can do it worse, or we can choose to pretend not to do it. But it will happen anyway.

An analogous case is, however, cautionary. Thanks to a poorly thought out, massive, and highly inefficient irrigation project intended to produce large amounts of poor-quality cotton, the Aral Sea in the old Soviet Union has in a few short years lost about half its area and about three-fourths of its volume. Diversion of almost 95 percent of its major feeder rivers, the Amu Dar'ya and Syr Dar'ya, has created a new 3-million-hectare desert, the Ak-kum. The spread of this desert has impacted the climates of China, India, and Southeastern Europe and driven twenty of twenty-four indigenous fish species extinct, destroying local fisheries and sixty thousand jobs in the process. It is devastation on a Promethean scale.

The first observation is that, while this case may be extreme, it is also archetypal; such hydrologic projects have been going on for centuries—indeed, they were foundational for Chinese civilization. And they are usually accompanied by unanticipated and undesirable impacts, and carry significant political overtones. So the Aral Sea example speaks to the history of human manipulation of major earth systems. But as political projects, such hydrological activities reflect specific historical and political contexts—in this case, the hubris of state Marxism—and thus are major mechanisms for building the contingency and reflexivity of human systems into natural systems. Large projects become a critical legitimating factor for any state, and in that guise they become avenues for what might well be called "rogue ESEM." Rogue ESEM can be thought of as ESEM which, for cultural, historical, and institutional reasons, escapes the rational, ethical, and theological checks and balances—and, importantly, the transparent and inclusive dialogues—that otherwise should inform ESEM projects. Like the Aral Sea, the result is frequently state-sponsored projects that destroy not just natural systems, but human ones as well. The hubris that lies behind such disasters is real, and an important danger to the practice of ESEM.

Perhaps the most obvious example of ESEM, however, is the global climate change negotiating process. Although it purports to respond only to a particular environmental perturbation, the implications of managing global climate change stretch across many human systems, simply because any climate management program will have huge economic, cultural, technological, and distributional effects. These impacts also extend across most natural systems from the nitrogen and hydrological cycles to the genetic system. In fact, it is doubtful that any meaningful management of climate can be achieved without much more direct management of carbon systems, and that will, in the opinion of many, require significant genetic bioengineering. It is not stretching the truth, but recognizing reality, to realize that climate change negotiations are nothing less than an attempt to engineer the future paths that civilization on this planet will have available for its evolution. Along these lines, remember that even the steps proposed in the Kyoto Protocol, significant enough to cause the United States to withdraw from the process, are almost immaterial compared to what will be required to reduce atmospheric carbon to the levels demanded by the environmental community. Given this responsibility, it is a powerful indictment of that community, and of environmental policy generally, that it is rejecting virtually any technology—nuclear power, biotechnology, geoengineering projects that would reduce the energy that the earth requires to begin with, and geophysical or oceanic carbon sequestration—that offer pathways to necessary carbon management systems. Although the dialogue around global climate change has been conducted in the language of science, it is an exercise in social engineering, with an almost religious set of teleological visions behind it. And this is also why, to a large extent, it has been relatively unsuccessful. Even if they are not conscious of the underlying dynamics, those who are to be socially engineered—predominantly the United States and its consumer society—tend to resent it.

The implications of ESEM, and the light it shines back on environmentalism and environmental policy, are thus quite challenging. The environmentalist discourse is generally highly imperialistic and sanctimonious, and dismissive of virtually any other interests; the ideologies of environmentalism dominate environmental science and reduce its objective validity; the teleologies of environmentalism are seldom explicated and even less seldom rationalized; and environmentalism has little patience with either dialogue or criticism, since it is self-evidently more moral and factually correct than any other discourse. It is thus perhaps ironic that the discourse that should be doing the most to inform the rational and ethical evolution of the anthropogenic world is, by choice and culture, least able to participate in such a dialogue constructively.

In the long term, however, this insularity and self-righteousness are probably, to a large extent, growing pains. As a discourse, environmentalism, which has always defined itself as countercultural, must now begin the difficult process of changing from adolescence to adulthood. The realities of the world—the very existence of significant anthropogenic environmental perturbations, the knowledge that human lives and values are as worthy of respect and support as nature, new foundational technologies like biotech, nanotech, and information technologies that are both desirable and unstoppable—will aid the maturation of environmentalism. New technologies in and of themselves will most likely have far more profound effects on the interrelationships between humans and environmental systems than all environmental policy put together.

As Thomas Kuhn pointed out years ago in *The Structure of Scientific Revolutions*, it will probably not be the environmentalists of today who can make that change, for they are too committed to past mental models and beliefs. We will need a new generation of nonenvironmental environmentalists if we are to achieve real progress. For our choice is not whether we want an anthropogenically influenced, engineered earth; that is already decided. Rather, our choice is whether to respond ethically and rationally to the challenge history has created for us.

For Further Exploration

Brad Allenby, Earth Systems Engineering and Management. *Technology and Society Magazine* (IEEE, Winter 2000/2001):10–24.

Thomas S. Kuhn, *The Structure of Scientific Revolutions*. University of Chicago Press, 1996.

Part III

New Governance

The cardinal tendency of progress is the replacement of an indifferent chance environment by a deliberately created one.

—J. D. Bernal, 1969

9

A Long Look Ahead: NGOs, Networks, and Future Social Evolution

David Ronfeldt

The information revolution favors the rise of network forms of organization, so much so that a coming age of networks will transform how societies are structured and interact. The environmental movement, like many other transnational civil-society movements, has gained strength over the past several decades largely because it has made increasing use of network forms of organization and strategy. In the years ahead, the movement's strength (and sometimes its weakness) will continue to be asserted through social network-based wars against unresponsive, misbehaving, or misguided corporate and governmental actors. But in the long term, network dynamics will enable policymakers, business leaders, and social activists to create new mechanisms for joint consultation, coordination, and cooperation spanning all levels of governance. Aging contentions that "the government" or "the market" is the solution to environmental or other particular public policy issues will give way to new ideas that "the network" is the optimal solution. The rise of network forms of organization and strategy will drive long-range social evolution in radical new directions.

Long-Range Social Evolution

Four forms of organization—and evidently only four—lie behind the evolution of all societies across the ages:

- The first to emerge and mature, beginning thousands of years ago, was the *tribal* form, which is ruled by kinship and clan dynamics and gives people a distinctive sense of identity, belonging, and culture, as manifested today in nationalism and even in fan clubs.
- The second to take shape was the *institutional* form, which emphasizes hierarchy and led to development of the state, as epitomized initially by the Roman Empire.
- The third to take hold is the *market* form, which excels at free and fair economic exchanges; it was present in ancient times but did not gain sway until the nineteenth century, starting mainly in England.
- The fourth to spread and mature is the *network* form, which is only now coming into its own, so far strengthening civil society more than other realms.

Each of the four forms, writ large, represents a distinctive set of beliefs, structures, and dynamics (with both bright and dark sides) of how a society should be organized—about who gets to achieve what, why, and how. Moreover, each form enables people to do something—to address some societal problem—better than they could by using another form. Each form engages different kinds of actors and adherents. Each has a different ideational and material basis, and of course, different organizational and leadership structures. Much of modern-day environmental policy and protection has drawn heavily on opportunities provided by the institutional and market forms. Today, collaborative networks are on the rise as the next great form, the cutting edge being among activist nongovernmental organizations (NGOs) associated with civil society.

As each form arises and matures, energizing a distinct set of values and norms for actors operating in that form, it generates a new realm of activity: the state, the market, and so forth. As a realm gains legitimacy and expands its space, it places new limits on the scope of existing realms. At the same time, through feedback and other interactions, the rise of a new form or realm also modifies the nature of the existing ones. An example would be the evolution of European absolutist regimes into democratic regimes, as the old hierarchical state institutions gave up on mercantilism and were remolded by the rise of the market system, including through the rise of marketlike electoral politics.

The main story is that societies advance by learning to use and combine all four forms, in a preferred progression. What ultimately matters is how the forms are added, and how well they function together. They are not substitutes for each other. Historically, a society's advance—its progress—depends on its ability to use all four forms and combine them (and their realms) into a coherent, balanced, well-functioning whole. Societies that can elevate the bright over the dark side of each form and achieve a new combination become more powerful and more capable of complex tasks than societies that do not. A society's leaders may try to deny or skip a form (as in the case of clannish ethnic groups that have trouble forming a real state, or Marxist-Leninist regimes that oppose the market), but any seeming success at this will prove temporary and limited. A society can also limit its prospects for evolutionary growth by elevating any one form to primacy—as appears to be a risk at times in market-mad America.

Over the course of time, the monoform societies organized in mainly tribal (T) terms—and many still exist today—are eventually surpassed by societies that develop institutional (I) systems to become biform T + I societies, often with strong states. These are in turn superseded by triform societies that allow space for the market (M), and become T + I + M societies. At present, the network (N) form is on the rise. Civil society appears to be the home realm for the network form, the realm that is being strengthened more than any other—but it is possible that a new, as yet unnamed realm will emerge from it. Thus a new phase of evolution is dawning in which quadriform T + I + M + N societies will emerge to take the lead, and a vast rebalancing of relations among state, market, and civil-societal actors will occur around the world. To do well in the twenty-first century, an advanced, democratic, information-age society must incorporate all four forms and make them function well together, despite their inherent contradictions.

In historical terms, it is often difficult—and it may take decades or longer—for a society to adapt to each new form and relate it to those that have developed earlier. Success is not inevitable. Every society, every culture, must move at its own pace and develop its own approach to each form and their combination. There is no single way to get a form or a combination right. What is right and wrong may vary from society to society. A society may get stuck, go astray, prefer to stay with an old design, or even be torn apart as it tries to adapt to a new form. For example, the great social revolutions of the twentieth century—in Mexico, Russia, China, and Cuba—occurred in mostly agrarian, biform T + I societies where old clannish and hierarchical structures came under enormous internal and external stresses that derived partly from inadequate or flawed infusions of capitalist market practices. Failing to make the +M transition, they reverted to hard-line T + I regimes that, except in Mexico, remolded absolutism into totalitarianism. Today, to varying degrees, these nations are trying anew to make the same + M transition. Again except for Mexico, none of them is yet hospitable to the presence of networked NGOs who represent + N dynamics.

The United States, along with countries in Western Europe and Scandinavia, is on the cutting edge of the world's current prospects for generating a quadriform T + I + M + N society, and this explains some of the social turbulence we have been experiencing at home and abroad. In general, the society that first succeeds at making a new combination stands to gain advantages over competitors and to have a paramount influence over the future nature of international conflict and cooperation. But if a major power finds itself stymied by the effort to achieve a new combination, it risks being superseded.

Implications for Future Modes of Conflict and Cooperation

New modes of conflict and cooperation emerge with each evolutionary shift. A society's efforts to transition from one stage to the next, or relate to a society that is at a different stage, are bound to create internal and external contradictions; indeed, the values, actors, and spaces favored by one form tend to contradict those favored by another. Thus, the rise of a new form induces epochal philosophical, ideological, and material struggles that are jarring to a society's stability, transformability, and sustainability. This happened in the past when tribal systems faced the rise of states, and states the rise of market systems. It will happen anew now that the network form is on the ascendance, energizing mainly nonstate actors.

Network forms of organization are attracting enormous attention these days; new books and articles appear every few months (with some of the best analysis coming from researchers at the Carnegie Endowment for International Peace). But it remains quite unclear what a new network realm may ultimately look like. One way it may evolve is through a four-stage progression in network topology: from an initial scattering of groups and individuals that have sparse network ties ("scattered emergence"); to their combining into a "single hub-and-spoke" design where the central hub acts mainly as a clearinghouse and coordinating agency; then to a deeper, more dispersed, specialized "multi-hub small world" design; and eventually to a dense, vast, sprawling "core/periphery" mass of organizational networks.[1] To some extent, the environmental movement (not to mention other civil-societal movements) is already moving through this progression. But just as the molding of a network realm will take time, so will the development and understanding of its interactions with the other, existing realms.

Near-Term Scenario: More Social Netwars

When a new form arises, it first has subversive effects on the old order, before it has additive effects that serve to consolidate a new order. For the age of networks, this subversive phase means the advent of netwar across the spectrum of conflict—from terrorist and criminal netwars at the violent end, to peaceable social netwars at the civil end.[2]

The term *netwar* refers to an emerging mode of conflict (and crime) at societal levels, short of traditional military warfare, in which the protagonists use network forms of organization and related doctrines, strategies, and technologies attuned to the information age. These protagonists are likely to consist of dispersed organizations, groups, and individuals who communicate, coordinate, and conduct their campaigns via the Internet, often without a precise central command. Thus, netwar differs from modes of conflict and crime in which the protagonists prefer to develop formal, stand-alone, hierarchical organizations, doctrines, and strategies, as in past efforts, for example, to build centralized movements along Leninist lines. In short, netwar is about Mexico's Zapatistas more than Cuba's Fidelistas, Hamas more than the Palestine Liberation Organization, the American Christian Patriot movement more than the Ku Klux Klan, and the Asian Triads more than the Cosa Nostra. In the "Battle of Seattle" the NGO activists who clogged the streets were the ones engaging in social netwar, not the traditional AFL-CIO marchers who held a largely ineffective mass meeting nearby.

Social netwar may be particularly effective where a set of protagonists engage in "swarming"—an approach to conflict that is quite different from traditional mass- and maneuver-based approaches. Swarming is a seemingly amorphous, but deliberately structured, coordinated, strategic way to strike from all directions at a particular point or points, by means of a sustainable pulsing of force, fire, or both from close-in as well as from stand-off positions. This "force and/or fire" may be literal in the case of military or police operations, but metaphorical in the case of NGO activists, who may, for example, be descending on city intersections or emitting volleys of e-mails and faxes. Swarming works best—perhaps it will only work—if it is designed mainly around the deployment of myriad, small, dispersed, networked maneuver units who converge on a target from multiple directions. The aim is sustainable pulsing—swarm networks must be able to coalesce rapidly and stealthily on a target, then sever and redisperse, immediately ready to recombine for a new pulse. Rapid information spread via Internet and communications systems may be indispensable for this to work well.

The Direct Action Network's operations in the Battle of Seattle provided an excellent example of swarming behavior that disrupted a major meeting of the World Trade Organization. Elsewhere, environmental groups such as Greenpeace and Robin Wood have used swarming to great effect in various environmental campaigns. Many of these actions have taken on David versus Goliath proportions as these NGOs have faced off against large transnational corporations or nation-state interests and won. Indeed, Home Depot, the world's largest lumber retailer, publicly committed itself to stop purchasing timber from endangered forests after the Rain Forest Action Network and Greenpeace organized an e-mail and mass media campaign based on Internet coordination among hundreds of environmental organizations and grassroots groups around the world.

Long-Term Scenario: New Approaches to Policy and Strategy

As the subversive effects subside and additive effects take hold, a society adapts to the rise of a new form and learns to combine it with the prevailing forms (realms). This may take decades, probably much longer. The deepening of the network age will cause a leap in the strength of civil society, or the emergence from it of a new network-based realm whose name and nature are not yet known. An end result will be the creation of next-generation policy mechanisms for communication, consultation, and collaboration among state, market, and civil-societal (or new-realm) actors, at home and abroad.

Because of the rise of a realm of networked actors and forces, current approaches to domestic and foreign policy will go through radical revisions. Some oft-noted trends will deepen: First, the boundaries between domestic and foreign policy will blur further, as activist NGOs continue to press for transnational perspectives on policy problems and solutions. Second, public-private cooperation, so needed in so many issue areas, will continue to extend beyond state and market actors to include socially minded nonprofit NGOs (who are sometimes said to comprise a new social sector separate from the traditional public and private sectors).

Meanwhile, new challenges will take shape. Taking advantage of the information revolution, people and organizations in advanced societies are developing vast sensory apparatuses for watching what is happening in their own societies and elsewhere around the world. Many innovations are occurring in organization and strategy, often by taking advantage of the new information and communications technologies. For example, one unusual benefactor of the Internet has been birdwatchers in North America, a group that contains an estimated one in four of all

U.S. citizens. A few years ago, with the help of the National Audubon Society and Cornell University, bird watchers started sharing data online and have used the network to better understand and map migration patterns.

Mining and analyzing data culled from large networks is not new and has long been used by existing government regulatory, law-enforcement, and intelligence agencies; corporate market-research departments; news media; and opinion-polling firms. What is new is the looming scope and scale of this sensory apparatus, and its increasing inclusion of NGOs who watch, monitor, share information, and report on what they see in diverse issue areas. One example is Global Forest Watch (GFW), an international network of local forest-protection groups linked by the Internet and a common data-gathering format. The World Resources Institute collaborated by e-mail with over a hundred scientists in different parts of the world to create a unique set of digital maps showing the location and extent of the world's old-growth forests. The GFW network monitors all these areas, recording any illegal cutting, burning, or other violations of forest leases on the digital maps. This information is posted to the Internet in near-real time, naming specific violators. Review processes check the accuracy of the data collected and ensure that participating network groups are acting responsibly. The information makes it possible for activists to mobilize quickly, apply market pressures to companies, and pressure governments to regulate effectively.

Developing the kind of early warning capability illustrated by GFW is an increasing concern for many environmental NGOs; so is gathering information to affect the framing of policy options. Determining appropriate designs, and roles, for this array of sensory organizations and their (centralized? decentralized?) internetting will be a growing challenge. Perhaps one day in the future, advances in autonomic and pervasive computing will even enable us to build self-regulating systems that can monitor and report on conditions without the constant involvement of humans. How such networked-based, self-regulating systems may interface with traditional forms of top-down regulation by government is only a speculative question today.

The emergence of a network realm—and of massively networked systems and infrastructures with it—will pose significant challenges for the agencies responsible for environmental protection. Their advisory councils and decision structures will have to open up to more regular participation from NGO representatives, at least in consultative capacities. Indeed, various environmental, health, consumer, and other activist watchdog and advocacy groups are already working—perhaps with more success in Canada and Europe than in the United States—to see that such reforms occur. And as stronger, more transparent connections are built among the responsible government agencies, NGOs, and companies, they will have to learn to work more cooperatively to formulate policy through new governance systems that embrace not only government and business but also NGO representatives. Climate change is probably the best example of an environmental issue that can only be resolved by networking across institutional boundaries of every kind.

Public policy dialogue has, for over a century, revolved around contentions as to whether government or the market represents the better solution for particular policy issues. In the network age, this choice will prove too narrow, too binary, even for blending. New views will come to the fore that the *network is the solution*. These views may well open up possibilities for major improvements in environmental protection.

Historically, the environmental community began by seeking top-down, command and control strategies (and still does) and then branched out in the 1980s into market-based incentives (emissions trading, etc.). Environmental NGOs have used network-based strategies to influence corporations and governments. But agencies responsible for environmental protection have only just begun to explore the potential of network designs for improving environmental monitoring and policy making and for accelerating progress in environmental science. Moving environmental protection into the age of networks is crucial for identifying and heading off potential environmental problems in their early stages, before they have massive impacts. It is also essential for dealing with the really hard environmental problems, the ones that require global cooperation and, the ones we have not dealt with yet because they are diffuse and distributed, involving myriad small sources with large aggregate impacts.

If this view of the role of networks—and of the eventual rise of a new network realm—in long-range social evolution is correct, the growth of transnational NGOs, and the ability of NGOs, states, and market actors to network with each other, should prove a major asset for democratic societies. It may help reanimate the concept of progress—giving it new direction and credibility. It may point the way to developing the structures and organizational processes that will make a sustainable future possible.

For Further Exploration

John Arquilla and David Ronfeldt, *Networks and Netwars: The Future of Terror, Crime, and Militancy*. Santa Monica, CA: RAND, MR-1382-OSD, 2001. Full text posted online at www.rand.org/publications/P/P7967/.

Krebs, Valdis, and Holley, "Building Sustainable Communities through Network Building," 2002. Posted at http://www.orgnet.com/Building Networks.pdf.

David Ronfeldt, *Tribes, Institutions, Markets, Networks: A Framework about Societal Evolution*, Santa Monica, CA: RAND, P-7967, 1996. Full text posted online at www.rand.org/publications/MR/MR1382/.

Notes

1. This four-stage progression is taken from a 2002 article by Krebs and Holley.
2. The concept of netwars is taken from John Arquilla and David Ronfeldt's *Networks and Netwars: The Future of Terror, Crime, and Militancy*. Santa Monica, CA: RAND, MR-1382-OSD, 2001. Full text posted online at www.rand.org/publications/P/P7967/.

10

Environmental Leadership in Government

Joanne B. Ciulla

People often ask, "Where have all the great leaders gone?" The usual answer is, "They're dead." In today's world, it is much easier to be a dead great leader than a live one. Leaders in all areas face the pressures of competing constituencies, shrinking resources, and a complexity of relationships and influences that give even local decisions a global context. Leaders have to be liked enough to get the job, but to do the job well, they must be prepared to be disliked and risk losing their jobs.

When it comes to U.S. environmental leadership the question is not "Where have the great ones gone?" but "Why, despite the best efforts of environmental groups, are there no great environmental leaders in government?" In Germany, the Green Party put environmental issues into the mainstream of politics. Here the environmental movement is still a social movement. It has not become a guiding force in government. Why? To answer this, we will examine some of the distinctive elements of environmental leadership.

The Problem of Power

Environmental leadership, like all other areas of leadership, requires some source or sources of power and influence. The traditional categories of power are transactional power (the power to reward and punish), the power of position, expert power, and personal power. To this I add a fifth source, moral authority. One reason why the environmental movement has been mostly a social movement is because its leaders do not have much transactional power, except within their organizations. Also the value system of our government does not give environmental leaders the same positional status as other leaders. The position of president of the Sierra Club does not have the same clout as the president of Exxon. (This goes back to transactional power.) The head of the Environmental Protection Agency (EPA) does not have the same clout as the Secretary of Defense. Despite years of discussions, the EPA still lacks cabinet status in the government, a visible symbol of its lower-class position within the executive branch. The power we attach to a position comes from the priorities of leaders and their followers. Government thrives on transactional and positional power. Leaders in government often see the environment as a luxury item we can only afford after we take care of the economy and defense—the icing, not the cake.

The second three sources of power—expert, personal, and moral—are perhaps the most important for today's environmental leaders. Unfortunately, they are often the least effective in government. Since the government is subject to a variety of constituencies, political leaders pick and choose the experts who best support their arguments. So, experts in government often have power because of their stance on an environmental issue (i.e., there is no clear evidence of global warming), rather than the actual quality of their expertise.

Personal power is very important for an environmental leader at the mid-manager level. It is the power he or she gets from being friendly, charismatic, tactful, charming, diplomatic, empathetic, and able to instill loyalty. One does not have to be charismatic to have personal power, but personal power is the basis of charismatic leadership. In government it is power that comes from managing people well and building an effective team. The head of the EPA may build a great team, but without other sources of power he or she may never achieve objectives with outside constituencies.

The last source of power is moral authority. People who are known for consistently telling the truth, keeping promises, fairness, and commitment to doing what they think is right, regardless of their self-interest, have a kind of power. It is a quiet power that seems to grow in its potency the more our society comes to mistrust both government and our leaders. Leaders with this kind of power instill trust. Trust is a powerful currency. It is not only useful for getting work done, but it is central to negotiation, conflict resolution, and coalition building. While personal power and charismatic power come from the likeability of or emotional attachment to a leader, moral authority comes from knowledge of a leader's track record and integrity. Such leaders never have to say, "You can trust me." Institutions and organizations should be set up to encourage leaders like this to emerge, but all too often in public administration and business, these people end up as either whistle-blowers, outsiders, or "pains in the neck."

The combination of personal power and moral power are hallmarks of *transformational leadership*. As James Macgregor Burns observes, these are the leaders who are willing to engage in dialogue and conflicts about values—and values are at the core of any transformation to a more sustainable society. Through this dialogue between various constituencies, transformational leadership strives to highlight and integrate the highest values of all parties concerned. Burns, ever the pragmatist, realized that transforming leaders also have to exercise transactional power to be effective. And herein lies the rub. How do we find, or grow, a new generation of environmental leaders with this complex mix of qualities and access to multiple power bases?

Moral and Scientific Arguments

Over the past two decades, coalitions and public-private partnerships have been instrumental in bringing about positive environmental change. What is less understood and appreciated is that members of these diverse coalitions draw their power from different sources. The sources of power that are not always effective in government are the central sources of power and influence in nonprofit and social movement leaders. These leaders gain power through their expert knowledge about the issues, personal skills, and moral authority. They need all of these to gain legitimacy, public attention, and funding, and to inspire volunteers.

The strengths of environmental leaders in the nonprofit sector can also be their weaknesses. Environmental issues entail both scientific and ethical questions about what is right and wrong in our relationships to people and all living things. Sometimes environmental leaders frame environmental issues in strong moral terms, but do not have the scientific background to make a compelling argument. There is a difference between the use of facts in a moral argument and their use in a scientific argument. While moral judgments rest on facts, they are not always decided by facts. We call something a "judgment" because we can never have all the facts but we are sure about the values and principles involved.

On the flip side are environmental leaders who make compelling scientific and economic arguments for issues such as why we need to reduce greenhouse gas emissions. This approach can result in a "if we could only get the facts right, then we would know what to do" mentality. The problem here is that when scientists disagree about the facts, people do not know what to do. Also, some scientists hold the rationalist view that assumes nature is an orderly system of patterns and principles that can be explained with mathematical models. The academic bickering that takes place between those who hold this point of view not only confuses the public, but it allows politicians and advocacy groups to pick whichever model best suits their purposes. Other scientists take a systems approach. For them the environment has a history, not a plan. They empirically chart this history by studying the causal connections of elements in complex systems. The systems approach is empirically based and hence messier for the scientist but easier for the public to understand. The long and short of it is that neither scientists engaged in technical battles nor marginalized individuals in social movements or in government have what it takes to create a clear vision of our obligations to the environment. But potential environmental leaders face a challenge that goes even deeper than trade-offs between scientific evidence and moral judgment and involves psychological traps built around beliefs in the righteousness of the cause and the institutional inbreeding that accompanies this stance.

The Trap of Self-Righteousness and Internal Focus

Those who are drawn to work in the environmental area are usually driven by a strong sense of the moral importance of their work. While this sense of commitment and urgency is what makes them successful in social movements and nonprofits, it can also produce leaders who are unwilling to compromise because they do not want to dilute the purity of themselves or their objectives. This works well for advocacy and public awareness, but as a model of leadership it is not very effective in producing constructive results when there are a number of different stakeholders.

Environmental groups often compete with each other for attention and funding. The commitment to the righteousness of a cause and need for the organization to survive can give an organization an internal focus. The same thing happens in government. Agencies compete for resources and their leaders turn inward, losing sight of the big picture. This lack of the big picture makes real transformational change difficult or impossible and blinds our leaders to both threats and opportunities that lie outside their picture frame. Just as scientists need to think systemically about the causal connections in the environment, environmental leaders need to think systemically about how to build partnerships with similar and very dissimilar groups that are causally connected to an issue.

In a recent study of environmental leaders, the majority of respondents identified interpersonal, technical, and conceptual skills as their most important leadership skills. It is striking that only one-quarter of the leaders surveyed identified political skills as important in their work.[1] One explanation is that they do not want to be sullied by politics. The other more distressing observation is that they do not think political skills are important. Either way this response makes sense, since their legitimacy rests on being "pure" in regard to their cause and playing politics is often regarded as a suspicious or dirty game. For example, Al Gore's book *Earth in the Balance* was not taken seriously by some academics *because* Gore was a politician. The flip side of this was that early on some of his political opponents tried to use his environmental concern as a means for marginalizing him and painting him as an extremist like the rest of the "tree huggers."

As vice president, Gore had positional, expert, and transactional power but he lacked the personal skills needed to create a transforming vision of the environment for government and the American public. His penchant for presenting his case with endless statistics and studies often invited sleep or bickering over the facts rather than initiating public and legislative dialogue about our values as a nation and as members of a global community. Gore's desire to educate the public about environmental questions was highly admirable, but he used the wrong lesson plan.

The Tall Order

We can now return to our early question: "Why, despite the best efforts of environmental groups, are there no great environmental leaders in government?" There are two related considerations for environmental leaders in public administration. The first involves the kinds of knowledge and skills needed. The second involves the strategies they need to be effective. The late John Gardner rightly observed that most great leaders have been generalists. He wrote:

> Tomorrow's leaders will, very likely, have begun life as trained specialists, but to mature as leaders they must crawl out of the trenches of specialization and rise above the boundaries that separate various segments of society. Young potential leaders must be able to see how whole systems function, and how interactions with neighboring systems may be constructively managed.[2]

A 1992 Conservation Fund survey of environmental leaders found that 46 percent had undergraduate degrees in the sciences, 25 percent had degrees in the social sciences, and 6 percent had degrees in technical fields. The rest of those surveyed had degrees in the humanities. Among those who held master's degrees, 60 percent had degrees in the sciences. The study concluded that these leaders have a wide variety of educational backgrounds.

While there is a variety in the content of these majors, there is not a variety in the epistemology of the sciences, social sciences, and technical areas. Despite claims to the contrary, many social sciences are taught with a value-free positivist slant. The humanities give students a broader worldview than the sciences or social sciences, but are frequently undervalued in the workplace. Organizations need to look more closely at the English, art, history, and philosophy majors for potential leaders. Environmental leaders of the future will need to look far beyond the environment for the solutions to tomorrow's problems. They will need the flexibility not only to journey into other disciplines but to wander into the interstitial spaces between disciplines where new innovations and the seeds of new coalitions may lie.

Perhaps the most important feature to look for in an environmental leader in government is a generalist who knows how to keep in perspective the work of the hardheaded specialists such as engineers, lawyers, and scientists who dominate the environmental field. How the next generation of environmental professionals is educated by our universities really matters and fostering leadership skills must be as important as inculcating an understanding of ecosystems or production processes. Few undergraduate or graduate programs in environmental management or science have even recognized this need, let alone developed the appropriate curriculum and training to address it.

A great environmental leader must be part poet, part scientist, part moral philosopher, and part politician. This is a tall order. We see some of these qualities in the work of the icons of the environmental movement, Aldo Leopold and Rachel Carson. Leopold and Carson made the case for preserving the environment with dazzling prose and inescapable logic. Their work animated and informed a moral vision of where we are and where we need to go. They were effective because they realized that the way we treat the environment is first and foremost a moral question about how we should live. Environmental leaders in government need to learn from what is best about nonprofit leaders and writers like Carson and Leopold. They should be advocates, communicators, and educators. To take on these functions, environmental leaders must create organizations that look outward and take a much more proactive stance toward the future. This means regarding other agencies and constituencies as potential allies, not competitors or enemies. By educating the public and taking their case directly to the public, environmental leaders will shape public opinion. By shaping public opinion, they will gain the transactional clout they need to do their work.

We need leaders in government who are willing to work to bring groups together and gain consensus on our moral obligations. When people agree on the values and moral principles it is usually much easier for them to hammer out the details of policy. This is something American leaders failed to grasp when they rejected the Kyoto Treaty. Rather than sign on to the *principle* of reducing greenhouse emissions for the good of the planet and work out their disagreements about its implementation, they chose to opt out of the process—a sad case of literally not seeing the forest for the trees.

We will also need leaders in government who can make hard decisions, often going against the political grain and organizational pressures to maintain the status quo. Sometimes top-down leadership is a good thing in a democracy, especially when the moral principles are so compelling. Some people believe that there are no great leaders because we don't live in a heroic era. Yet, when it comes to the environment, leaders in government have a unique opportunity for heroism. This is not the glamorous heroism with a cape, it's heroism with a cost. People and businesses will have to change, sacrifice, and spend money *and* they will not always see immediate results from their efforts. To do this environmental leaders will need to have the imagination and eloquence to animate a viable moral vision. As some of the authors in this book have written, people are beginning to coalesce around a vision of a more sustainable planet, built on transformational technologies and policies. These critical transformations are unlikely to happen without transformational leadership. This kind of transforming leadership takes enormous skill and personal integrity. It also requires the moral courage to risk one's career to do what is right. Environmental leaders on the outside of government have gotten us off to a start, but only through strong environmental leadership in government will we get the job done.

For Further Exploration

J. M. Burns, *Leadership*. New York: The Free Press, 1978.

J. B. Ciulla, *The Ethics of Leadership*. Belmont, CA: Wadsworth, 2003.

J. B. Ciulla (Ed.), *Ethics, the Heart of Leadership*. Westport, CT: Praeger, 1998.

C. P. Egri and S. Herman, Leadership in the North American Environmental Sector: Values, Leadership Styles, and Contexts of Environmental Leaders and Their Organizations. *Academy of Management Journal* 45(4, 2000):571–604.

D. Ehrenfeld, *Beginning Again: People and Nature in the New Millennium*. Oxford: Oxford University Press, 1993.

J. Gardner, *On Leadership*. New York: The Free Press, 1990.

A. Gore, *Earth in the Balance: Ecology and the Human Spirit*. Boston: Houghton Mifflin Co., 1992.

M. K. Landy, M. J. Roberts, and S. R. Thomas, *The Environmental Protection Agency: Asking the Wrong Questions from Nixon to Clinton*. Oxford: Oxford University Press, 1994.

A. Leopold, *A Sand County Almanac*. Oxford: Oxford University Press, 1989.

J. O'Neill, *Ecology, Policy and Politics*. New York: Routledge, 1993.

D. Pimentel, L. Westra, and R. F. Noss, *Ecological Integrity: Integrating Environment, Conservation, and Health*. Washington, D.C.: Island Press, 2000.

D. Snow, *Inside the Environmental Movement: Meeting the Leadership Challenge*. Washington, D.C.: Island Press, 1992.

Notes

1. Values, Leadership Styles, and Contexts of Environmental Leaders and Their Organizations. *Academy of Management Journal* 45 (4, 2000):571–604.
2. J. Gardner, *On Leadership*. New York: The Free Press, 1990.

11

Time Matters

Stewart Brand

It didn't start as a law; it started as a prediction. In retrospect it turned out to be the most accurate and consequential prediction in the history of technology, and it exposed the structure of technological hyperacceleration in the late twentieth century. The prediction appeared in the April 19, 1965, issue of the technical journal *Electronics* in one of a series of papers called "The Experts Look Ahead." The author was an electrical engineer with a background in physical chemistry. The title of the article was "Cramming More Components onto Integrated Circuits." The author's name was Gordon Moore, later cofounder of Intel Corporation, the world's leading computer chip manufacturer.

What became known as Moore's Law was a small graph and explanation buried in his *Electronics* paper. From 1965, Moore looked back to the beginnings of integrated circuits in 1959 and noted that the number of components that could fit on a chip had doubled every year for six years. He predicted that the trend would continue for at least another ten years, permitting an astonishing sixty-five thousand components on a chip by 1975. The actual numbers by 1975 were around twelve thousand, so the formula was later adjusted to predict a doubling every eighteen months.

History veered—not only as a result of the power of the new technology of computation, communication, and intelligence but also owing to the self-accelerating rate of its arrival described by Moore's Law. Dense computer chips were used to design still denser computer chips, ad infinitum. Doubling the number of components on a chip every eighteen months kept doubling computer power and halving expense. The explosion burst past 1975, continued through 1985, 1995, and it shows every sign of constant acceleration through at least 2015: thirty-seven doublings, about a 137-billion-fold increase in power in fifty-six years. There is no precedent in the history of technology for the sustained self-feeding growth of computer capability.

The pace of Moore's Law has become the pacesetter for human events. Velocity itself has become the dominating characteristic of the world's quicksilver economy. "We are moving from a world in which the big eat the small," remarked Klaus Schwab, head of the World Economic Forum, "to a world where the fast eat the slow." Technological acceleration is driving economic acceleration, globalization, and the intensification of global economic competition.

Civilization is revving itself into a pathologically short attention span. What with accelerating technology and the next-quarter perspective that goes with electronically accelerated market economies and the next-election perspective that goes with the spread of democracy, we have a situation where steady but gradual environmental degradation escapes our notice. Preoccupied with breaking news, we risk falling victim to slow problems.

Slow and steady events have often been fateful in history. Edward Gibbon's *Decline and Fall of the Roman Empire* excelled in this kind of perception; it's right there in the title. For Gibbon, writes Robert D. Kaplan,

> The more gradual and hidden the change, the more important it turned out to be.... The real changes were insidious transformations: Rome moving from democracy to the trappings of democracy to military rule; Milan in Italy and Nicomedia in Asia Minor beginning to function as capital cities decades before the formal division of the empire into western and eastern halves,

> and almost two centuries before Rome ceased to be an imperial capital. . . .

The sociologist Elise Boulding diagnosed the problem of our time as "temporal exhaustion": "If one is mentally out of breath all the time from dealing with the present, there is no energy left for imagining the future."[1]

In a 1978 paper, Boulding proposed a simple solution: expand our idea of the present to two hundred years—a hundred years forward, a hundred years back. A personally experienceable, generations-based period of time, it reaches from grandparents to grandchildren—people to whom we feel responsible. Boulding, a mother of five, wrote that a two-hundred-year present "will not make us prophets or seers, but it will give us an at-homeness with our changing times comparable to that which parents can have with an ever-changing family of children as they move from age to age."[2]

A two-hundred-year present is good; there is emotional comfort and behavioral discipline in it. If we want a truly profound change in mindset, however, even two hundred years may be too incremental. Frames of mind change by jumps, not by degrees. Ten thousand years is the size of civilization thus far. In that time a number of civilizations and dozens of empires have risen and fallen or receded, but the overall advance and convergence of civilization on the planet has been steady. A ten-thousand-year perspective places us where we belong, neither at the end of history nor at the beginning, but in the middle of civilization's story. It makes a two-hundred-year present seem homey indeed.

What sort of mechanism or myth could encourage the long view and the taking of long-term responsibility, where "the long term" is measured at least in centuries? What some of us propose is both a mechanism and a myth. It began with an observation and idea by computer designer Daniel Hillis, who wrote in 1993:

> When I was a child, people used to talk about what would happen in the year 2000. Now, thirty years later, they still talk about what will happen by the year 2000. The future has been shrinking by one year per year for my entire life.

> I think it time for us to start a long-term project that gets people thinking past the mental barrier of the Millennium. I would like to propose a large (think Stonehenge) mechanical clock . . . It ticks once a year, bongs once a century, and the cuckoo comes out every millennium.

Hillis, who developed the "massive parallel" architecture of the current generation of supercomputers, is now building the prototype of a clock designed to run for ten thousand years. Its works consist of an ingenious binary digital-mechanical system that has precision equal to one day in twenty thousand years, and it self-corrects by phase locking to the noon sun. In a 1994 discussion about how to name Danny Hillis's clock, ambient music pioneer Brian Eno suggested, "How about calling it 'The Clock of the Long Now,' since the idea is to extend our concept of the present in both directions, making the present longer. Civilizations with long nows look after things better."

A Clock of the Long Now, if sufficiently impressive and well engineered, would embody deep time for people. It would be charismatic to visit, interesting to think about, and famous enough to become iconic in public discourse. Ideally, it would do for thinking about time what the photographs of earth from space have done for thinking about the environment. It turned out that the astronomer Fred Hoyle was right in 1947 when he forecast, "Once a photograph of the Earth, taken from outside, is available . . . a new idea, as powerful as any in history, will be let loose."

The environmentalist Rene Dubos was also right: "We are becoming planetized probably almost as fast as the planet is becoming humanized."[3] Our global influence and our global perspective are almost keeping pace with each other, which is fortunate—it could have been otherwise. Once we acknowledge our new responsibility for the health of the planet, the large view and the long view become one. The Big Here and the Long Now merge as the Long Here, which is no longer just occupied but managed by what might be called the Long Us. The Chinese have a term for it: *da wo*, or "big me."

What would it mean to operationalize such a perspective? I believe it would help us see that significantly different levels of pace need to be recognized and maintained in the working structure of a robust and adaptable civilization. From fast to slow the levels are:

- Fashion/art
- Commerce
- Infrastructure
- Governance
- Culture
- Nature

In a healthy society each level is allowed to operate at its own pace, safely sustained by the slower levels below and kept invigorated by the livelier levels above. Each level must respect the different pace of the others. If commerce, for example, is allowed by governance and culture to push nature at a commercial pace, all-supporting natural forests, fisheries, and aquifers will be lost. If governance is changed too suddenly, you get the catastrophic French and Russian revolutions.

For governance to play its most effective role, it needs to respond to changes in the faster levels above it, yet maintain its place in the hierarchy of levels and make full use of the tremendously powerful lever of time. The full power of time has seldom been employed. The pyramids of Egypt and Central America took only fifty years to build. Some of the great cathedrals of Europe were built over centuries, but that was due to funding problems rather than patience. Humanity's heroic goals generally have been sought through quick, spectacular action ("We will land a man on the Moon in this decade") instead of a sustained accumulation of smaller, distributed efforts that might have overwhelming effect over time. The kinds of goals that can be reached quickly are rather limited, and work on them displaces attention and effort that might be spent on worthier, long-term goals.

Danny Hillis points out, "There are problems that are impossible if you think about them in two-years terms—which everyone does—but they're easy if you think in fifty-year terms." This category of problems includes nearly all the great ones of our time: the growing disparities between haves and have-nots, widespread hunger, dwindling freshwater resources, global climate change, loss of biodiversity, ethnic conflict, global organized crime, and so on. Such problems were slow to arrive, and they can only be solved at their own pace.

It is the job of slow and steady governance to set the goals of solving these problems and to maintain the constancy and patience required to see them through. That is not our current model of governance, but it is the mature character of governance we need to achieve. The maturity we need is more a matter of mindset and culture than organizational structure.

Maturity is largely a combination of hard-earned savvy, the habit of thinking ahead, and the patience to see long-term projects through. The embrace of duration yields wisdom, described by the scientist Jonas Salk as "the capability of making retrospective judgments prospectively." Wisdom decides forward as if back. Rather than making detailed, brittle plans for the future, wisdom puts its efforts into expanding general adaptive options. An earth with an intact ozone layer has more options than one without. A world rich in biodiversity has more potential than one without.

It is not easy sustaining endeavor to achieve long-term goals. Yet this is part of the attraction, that the task is nearly impossible seeming and bracingly hard. The rewards of immersion in a project, a story, reaching well beyond the span of one's own life, can be enormous. This is one of the things that keeps people working gladly in long-lived institutions such as universities and religions.

Environmental projects, owing to the extended lag times involved and perhaps the aesthetic rewards along the way, excel at inspiring long-term ambition. I know of two North American environmental projects with thousand-year time frames: Ecotrust, which is setting about building a nature-sustaining economy throughout the coastal temperate rain forest from mid-California to Northern Alaska; and the Wildlands Project, which aims to restore enough wild land, surrounded by partially wild "buffer zones" and connected by wildlife corridors, for native animal and plant populations to survive indefinitely amid human dominance of the continent. Instead of saving species individually and temporarily, the idea is to take the time to save them all permanently.

Government needs to set environmental goals on a similar scale, based on our collective aspirations for a sustainable future. Then it needs to set milestones to be achieved along the way. Then it needs to persevere.

The learning theorist Seymour Papert tells of a group of friends eating lobsters at a Boston fish house. The question came up, "Can anyone eat a lobster without making a mess?" Papert reports, "A brain surgeon at the table did it. It took him two hours—a completely eaten lobster with a perfect absence of mess. He took the time appropriate to the job, which he knew about. It wasn't his skill. It was his patience."

Two hours was the difference between impossible and easy. For what tasks would two hundred years make that kind of difference?

For Further Exploration

Stewart Brand, *The Clock of the Long Now*. New York: Basic Books, 1999.

Stewart Brand's web site: www.well.com/user/sbb/.

The Long Now Foundation web site: www.longnow.org/.

Notes

1. Elise Boulding, "The Dynamics of Imagining Futures," *World Future Society Bulletin,* September 1978, p. 7.
2. *Ibid.*
3. René Dubos, *The Wooing of the Earth,* New York: Scribners, 1980, p. 70.

12

The Guardian Reborn: A New Government Role in Environmental Protection

William McDonough and Michael Braungart

For more than thirty years, since the passage of the National Environmental Policy Act (NEPA) in 1969, environmental protection has been nearly synonymous with environmental regulation. During the century before the 1960s, in the spirit of outdoorsmen like John Muir and Teddy Roosevelt, protecting the environment almost always meant setting aside land for national parks. Rachel Carson changed all that. Her sober indictment of the chemical industry, *Silent Spring*, offered a startling new paradigm. The impact of industry, long regarded as limited to the visible effects of harvesting natural resources and generating waste, was far more widespread, more pernicious, and more enduring than anyone had imagined. Indeed, the poisoning of the earth was unprecedented. Industrial chemicals were not only polluting the air, water, and soil, they were changing the very structure of our cells. By Carson's lights, Americans had every right to expect that the laws of the land would protect them from such harm. She put it this way: "If the Bill of Rights contains no guarantee that a citizen shall be secure against lethal poisons distributed either by private individuals or by public officials, it is surely only because our forefathers, despite their considerable wisdom and foresight, could conceive of no such problem."[1]

Yet there it was. And after the publication of *Silent Spring* in 1962, it was a problem millions of people were reading and worrying about—and demanding that government address. In less than a decade, Carson's critique was so well known, and the nascent environmental movement so influential, Congress passed in quick succession the Clean Water Act, the Clean Air Act, and NEPA, while President Nixon moved to establish the Environmental Protection Agency (EPA) to develop and enforce environmental laws.

Rightly so. One of the proper roles of government, along with shaping just principles in support of social and political life, is to act as a guardian of its citizens. Against great odds, a generation of EPA officials has performed that role with an energetic commitment to the public good. Reinforcing the agency's best work, nongovernmental organizations—such as Environmental Defense, Natural Resources Defense Council, and Greenpeace—have led a thirty-year public dialogue that has firmly established Americans' right to clean air, water, and soil. Environmental legislation, when it has been given sharp enough teeth, has enabled the agency to enforce the basic standards that fulfill that right. Without the Clean Water Act, Ohio's Cuyahoga River might still be in flames.

We have reached an impasse, however. The regulatory infrastructure, as much good as it has done, is not enough to effectively protect the environment. Water quality, for example, remains a pressing issue. Sediments and microorganisms not covered by the Clean Water Act continue to pollute 44 percent of U.S. waters. When polluting substances *are* regulated, that doesn't always lead to the remediation of environmental harm, a problem illustrated by the ongoing twenty-year battle between the EPA and General Electric over the cleanup of polychlorinated biphenyls (PCBs) in the Hudson River. If, under current conditions, protecting environmental health has proven so difficult, how will regulations deal with a projected fivefold increase in economic activity over the next fifty years?

International environmental regulations are also difficult to enforce. The Basel Convention, for example, is a United Nations treaty adopted to stop the flow of hazardous waste from leading industrial nations to developing countries. The United States has not ratified the convention, however, so U.S. companies are not bound by it. Consequently, the hazardous wastes from computers collected for "recycling" in Houston, Omaha, or Portland are being processed all over the world. At one documented site along the Lianjiang River, northeast of Hong Kong, hundreds of poor, migrant workers burn hazardous materials in the open air and run toxic, riverside acid works to recover precious metals from old computers.

The international regulatory landscape is not entirely bleak. Environmental initiatives in the European Union (EU), for example, suggest how some regulations can drive innovation and economic growth. On the heels of Chancellor Willy Brandt, who back in Carson's day declared that he wanted to see clear, blue sky over Germany's heavily industrialized Ruhr District, European politicians have understood the necessity of strong regulations, international cooperation, and a more proactive role for government and industry. Regulations that successfully curbed acid rain, for example, also created opportunities to use recovered sulfur emissions to make gypsum. Today, the EU's End of Life Vehicle Directive, which requires automakers to take full responsibility for the materials in the cars they produce, is stimulating innovations in design and manufacturing that will allow companies to effectively recover valuable resources through auto recycling. Similarly, the fast-growing environmental technology industry offers economic opportunity during this transitional step toward clean, healthy commerce.

Yet, even when corporations willingly cooperate with regulations—in Europe or the United States—industrial production can still harm people and the environment. Consider the EPA's annual Toxics Release Inventory (TRI). Established in 1986, the TRI gathers data from industrial facilities, which are required to report on the release of hazardous chemicals, as well as the location and quantities of chemicals stored on-site. The reporting is designed to notify nearby communities of possible public health problems. While the most recent TRI data shows that chemical releases have decreased roughly 48 percent since 1988, industrial facilities in 2000 released *7.1 billion pounds* of toxic substances, including persistent bio-accumulative chemicals, such as dioxins, mercury, and PCBs. A separate EPA report, released just weeks after the current TRI in June 2002, declared that 20 million Americans live in areas where elevated levels of toxic chemicals pose a cancer risk one hundred times greater than the levels at which EPA pollution-reduction programs typically target cancer-risk sources.

As evidence mounts that even tiny amounts of dangerous emissions can have harmful effects on biological systems over time, it seems prudent, if not urgent, to add some new options to the repertoires of both the guardian and the business leader—and even build cooperative relationships between them. Considering the traditional enmity between commercial interests and regulators—as in General Electric versus EPA—that might be hard to imagine. But it's not impossible. While many corporations still see regulations as obstacles to profitability and spend undue energy looking for loopholes to protect the bottom line, others are making environmental responsibility an integral part of their business agenda—and benefiting from doing so. When seen as a driver of quality and innovation, environmental concern can be quite profitable. In fact, many leading corporations are discovering that products, services, and manufacturing facilities can be designed to *enhance* environmental health, economic value, and quality of life. As you'll see, this is not a pipe dream. It's the emerging future of American economic strength, a future in which the EPA could play a vital role.

The first step might be a commitment to environmental protection that begins not with aiming to reduce the release of dangerous chemicals but attempting to eliminate toxic emissions altogether—by design.

From our perspective, regulation is *a signal of design failure*. The government's need to step in and manage toxic emissions does not suggest bad intentions on the part of either the guardians of the public realm or commercial interests. Rather, the conflict between nature and commerce, and between business and nature's protectors, is the result of deep flaws in the design of our current industrial system.

Traditional manufacturing creates such a bevy of negative consequences because it is built on a cradle-to-grave model that generates products designed for a one-way trip to the landfill and incinerator. The World Resources Institute (WRI) estimates that "one-half to three-quarters of annual resource inputs to industrial economies are returned to the environment as wastes within one year."[2] Attempts to limit manufacturing waste tend to fine-tune the engines of industry, diluting pollution and slowing the loss of natural resources without examining the design flaws at their source. Reforms such as these take for granted the antagonism between nature and industry. The result: business strategies and a regulatory environment built on restricting industry and curtailing growth—a dispiriting commercial and environmental dead end.

But what if our designs were so inherently productive and healthful they allowed us to celebrate the things we make? A strategy we call "cradle-to-cradle design" allows us to do so. It rejects the assumption that the natural world is inevitably destroyed by human industry, or that excessive demand for goods and services is the inevitable cause of environmental problems. Conventional industrial design is flawed because it developed in a time when few understood the dynamic relationship between economy and ecology, or the principles of the earth's natural systems.

The principles of cradle-to-cradle design, on the other hand, are modeled on the earth's natural systems, the perpetual flows of energy and nutrients that support biodiversity. The intention: to apply the intelligence and effectiveness of these systems to product, process, and facility design.

From an industrial design perspective, this means developing a deep understanding of the chemistry of materials and applying it to product development. It means creating supply chains and manufacturing processes that replace industry's cradle-to-grave model with systems modeled on nature's cradle-to-cradle cycles, in which one organism's waste becomes food for another. When designers and engineers apply these principles to product conception and material flow management, they can begin to create goods that flow effectively within closed-loop systems, providing after each useful life either nourishment for nature or high-quality materials for new products. Ultimately, we think cradle-to-cradle design can lay the foundation for an industrial system that restores nature, eliminates the concept of waste, and creates enduring wealth and social value—human industry as a regenerative force.

Put another way, we are offering a complement—and ultimately an alternative—to environmental regulation. Consider, for example, how cradle-to-cradle thinking, applied to the design of an industrial facility, can create a profitable outcome for both business and nature—one that would have never been imagined if approached from either a purely economic or a purely regulatory perspective.

In May 1999, Ford Motor Company decided to invest $2 billion over twenty years to restore its Rouge River manufacturing plant in Dearborn, Michigan. Working with Ford's executives, engineers, and designers, we began to explore innovative ways to rebuild the complex. Rather than using economic metrics to reconcile the apparent conflicts between environmental concerns and the bottom line, we set out to restore the Rouge to a life-supporting place.

The systems for storm water management were one of our key concerns. Typically, expensive technical controls are the response to storm water regulations. Following industrial protocol, Ford had estimated that new pipes and treatment plants would cost up to $48 million. If we had approached the flow of water on the Rouge from an economic or regulatory perspective we might have tried to cut costs by using pipes made with less material, or by finding ways to treat water with fewer harmful chemicals.

Instead, we designed the plant to create habitat, make oxygen, connect employees to their surroundings, and invite the return of native species. The result, now under construction, is a daylit factory with 450,000 square feet of roof covered with healthy topsoil and growing plants—a living roof. Along with porous paving and a series of constructed swales, the living roof will slow and filter stormwater runoff, making expensive technical controls, and even regulations, unnecessary. All this with first cost savings of up to $35 million, with the landscape thrown in for free.

This is clearly an example of the carrot being far more compelling than the stick. Wouldn't it be marvelous if the EPA could create a new relationship with commerce that allowed designs such as these to emerge and evolve throughout American industry? Imagine the EPA offering incredibly sweet carrots to industries hungry for new ideas. Imagine an agency supporting innovative, ecologically intelligent designs. Developing cradle-to-cradle benchmarks for materials, products, and facilities and presenting them to industry as practical, productive strategies that effectively protect, and even restore, the environment.

This is not at all far-fetched. The EPA is already developing proactive projects and reaching out to industry through its Green Chemistry, Design for the Environment, and Product Stewardship programs. Admittedly, these are small efforts in the grand scheme of things, but they show that the EPA has something it can grow.

"While keeping current regulations in place, we are going to have to develop some new tools," said Derry Allen, counselor to the EPA Office of Environmental Policy Innovation. "When you look ahead a generation, the future of the EPA has to include thinking about new strategies for material use and product design."

When sufficient energy develops within the agency to pursue a new path, the benchmarks are out there to be studied and presented to industry. Ford's Rouge River plant is only one of many examples of inherently healthful, productive designs that are extremely attractive from the perspective of both the guardian and the business executive. The EPA and other government agencies could encourage designs such as these, supporting industry with information and know-how as the United States becomes a world leader in intelligent design and resource recovery. The result: a healthy environment, a growing economy, and a better quality of life for all Americans—and for the rest of the world.

This is a crucial step for American business. In recent years, as trade has rendered the boundaries between nations more fluid, American manufacturing has undergone a transformation. Corporations bent on achieving global reach have increasingly moved manufacturing operations overseas to nations that provide cheap labor and a less strict regulatory environment. This has proved to be a double-edged sword. While many businesses see their bottom line growing, they are increasingly reliant on factories and supply chains they do not own or manage. Consequently, few products are completely produced in the United States and few American companies know what is in their products—consumers and regulators don't know either. The international recycling of computers is just one example of how toxic products are made offshore, used by U.S. consumers and then shipped back overseas, creating a toxic flow of liabilities.

We need to reinvent our global business strategy. We need to redesign our manufacturing model so we can offer the world a system built on product quality, on design protocols founded on a thorough understanding of the chemistry, the value, and the beneficial effects of industrial materials. If we begin now to develop our commercial industries around cradle-to-cradle protocols, the United States can become the world leader in high-quality product design, rather than competing on uneven and unhealthy terms within the old industrial system. This would not only protect the health and well-being of American consumers, it would nourish the American economy and the American land. It would also yield exceedingly smart, effective benchmarks to export to developing nations, rather than exporting harm. And as we renew product quality, we will also be developing an intellectual infrastructure supporting the making of things that will give us long-term prosperity rather than short-term gain.

What an interesting irony that the U.S. EPA, so long considered the bane of business sense and productivity, just might be positioning itself to support this bold environmentally sound vision of American economic strength.

For Further Exploration

William McDonough and Michael Braungart, *Cradle to Cradle*. New York: North Point Press, 2002.

McDonough Braungart Design Chemistry web site: www.mbdc.com/.

Notes

1. Rachel Carson, *Silent Spring*, Marine Books, 2002.
2. Allen Hammond et al., *Resource Flows: The Material Basis of Industrial Economies*. Washington, D.C.: World Resources Institute, 1997.

13

Advancing Corporate Sustainability: A Critical New Role for Government

David V. J. Bell

Many governments around the world are beginning to recognize the importance of addressing the challenge of sustainability. It has become increasingly evident, however, that governments acting alone cannot achieve the far-reaching technological, social, and economic advances that sustainability will require. Though sustainable development began as a project for governments (in the report of the Brundtland Commission and the organization of the first Earth Summit in 1992), the need to engage all sectors of society is now self-evident.

Because the private sector controls such large resources and plays the primary role in converting new knowledge into new technologies and products, business must be a major part of the sustainability solution. Indeed, in a global market economy, business may be the key force for moving toward a sustainable future. As Ray Anderson, the CEO of Interface, Inc. argues, "only business ... the largest institution on Earth ... can lead [toward sustainability] quickly and effectively."

From this perspective, the most important emerging development that will shape the environmental future may be the appearance of a growing number of corporations that are striving to be *sustainable enterprises*—enterprises that create value while consciously avoiding damage to economic, social, or natural capital, and that operate on principles of transparency and accountability.

Advancing corporate sustainability is one of the most important roles government can play over the generation ahead. Government has an enormous opportunity to leverage constructive change by facilitating a business transition to an economy that is much more efficient, much more fair, and much less damaging. But to take advantage of this opportunity, government itself will need to innovate.

The Business Case for Sustainable Enterprise

A growing number of leading businesses are pursuing sustainable development because they see it as a value proposition and are able to make a strong business case for doing so. They believe enormous opportunities will open up over the years ahead for those companies that are well positioned to move toward the high technology-based sustainable economy of the future. And they believe the short-term advantages of moving toward sustainability are increasingly compelling.

The World Business Council for Sustainable Development (WBCSD), a coalition of approximately 150 international companies, recently prepared an analysis of *The Business Case for Sustainable Development*. It concludes that "pursuing a mission of sustainable development can make our firms more competitive, more resilient to shocks, nimbler in a fast-changing world, more unified in purpose, more likely to attract and hold customers and the best employees, and more at ease with regulators, banks, insurers, and financial markets."

The potential consequences of *not* pursuing sustainable development are a less immediate but still significant motivating factor. There are real risks that economic development around the world can be slowed or derailed by environmental problems and social disorder. Business strategies that simultaneously create economic value, protect the environment, improve security, and reduce poverty and inequity can minimize these risks. The ultimate bottom line is that business cannot succeed in societies that fail.

Key Roles of Government in Promoting Sustainable Enterprise

There is a great deal that governments can do to promote sustainable enterprise. No fundamental changes in the role of government are needed, only a systematic effort to redirect existing roles of government toward promoting sustainable development.

Visionary and Goal Setter

The metaphors "steering" and "rowing" may help distinguish the respective roles of government and business. Business is best equipped to generate wealth by providing products and services. As rowing moves a boat, so business activity is the force that propels a healthy economy. Government's role, by contrast, is to steer society toward goals that are articulated in public policy.

Several governments have already begun to articulate a vision and strategies to help guide policy in the direction of sustainability. For example the Organisation for Economic Co-operation and Development (OECD) Environmental Strategy for the First Decade of the 21st Century is intended to provide clear directions for environmentally sustainable policies in OECD member countries. The Strategy is to be fully implemented by 2010. OECD Environmental Performance Reviews and the Environmental Indicators Programme will be used to monitor progress.

Still more comprehensive visions are needed of the character of an ecological economy, an environmentally advanced technology, and a sustainable society. Wide-ranging public discussions are needed to work toward greater consensus on goals for the transformations that will be needed in social and economic systems dealing with energy, waste, water, production, consumption, social inclusion, trade, and so on.

Collaborator and Partner

A shift in focus is underway from *government* to *governance*. Government refers to particular kinds of public institutions (the "state") that are vested with formal authority to take decisions on behalf of the entire community. Governance encompasses collective decisions made in the public sector, the private sector, and civil society. It suggests the need for collaboration among these sectors to address the kinds of broad challenges associated with sustainability.

Collaboration for sustainability means that increasingly governments must form partnerships—with other levels of government, with the private sector, and with civil society organizations. This imperative creates both dangers and opportunities. The danger is that government will fail to recognize its distinct obligations within such partnerships or will choose this approach even when it is inappropriate. The opportunity is to extend the commitment to sustainability throughout society and to combine skills and provide access to constituencies that no one partner may have. Partnerships also enhance the credibility of results—results that might be less effective and believable if they come from only business, or only civil society, or only government.

Leader by Example

In most countries, the government is the largest landowner, the largest fleet owner, the largest single employer, and the largest landlord or owner and operator of buildings. It is also therefore the largest consumer of energy, the largest producer of most environmental impacts, and the greatest single source of support for social capital. A strong case can be made that governments should "walk the talk" by putting their internal operations on a firm sustainability foundation. Just as most governments try to conduct government operations and public enterprise according to sound business practices, sustainability principles should now be seen as integral to this process.

Japan's successful use of procurement to stimulate technological innovation in fuel-electric hybrids and other low-emission vehicles illustrates the benefits this strategy can provide. It has contributed to better environmental performance, increased international competitiveness, and lowered consumer prices for these vehicles. This initiative has broken the vicious circle that surrounds many innovative environmental technologies: low (initial) demand means less production and therefore higher prices.

Facilitator

Governments can accelerate the shift to sustainable enterprise by creating open, competitive, and rightly framed markets. The World Business Council for Sustainable Development (WBCSD) recommends that governments initiate a steady, predictable, negotiated move toward:

- Greater use of market instruments and less of command and control regulations
- Dismantling perverse subsidies
- Full-cost pricing of natural resources, goods, and services
- Fiscally neutral tax changes to reduce taxes on things to be encouraged (work, saving, investment) and to raise taxes on things to be discouraged (waste, pollution)
- More reflection of environmental resource use in Standard National Accounts

Every OECD country has adopted one or more of these elements, but no government has adopted this full set of framework conditions as national policy.

Green Fiscal Authority

Since Rio, research on "green budgeting" and ecological fiscal reform (EFR) has expanded enormously. A recent OECD report entitled *Environmentally Related Taxes in OECD Countries: Issues and Strategies* reports that all OECD countries have introduced environmental taxes to a varying extent. An increasing number of countries are implementing comprehensive green-tax reforms, while others are contemplating doing so. The report identifies obstacles to a broader use of such taxes—in particular the fear of loss of sectoral competitiveness—and ways to overcome such problems.

The WBCSD strongly supports greater reliance on market solutions, pointing out they are not only "among the most powerful tools available, but—properly structured—they can be among the least painful." Noting that the market is "not good at pricing many environmental assets and services like a stable climate or rich biodiversity," the WBCSD recommends approaches such as tradable permit schemes and other efforts to "create a market" by assigning monetary values to natural resources and natural services. "We do not protect what we do not value. . . . Establishing such prices—in ways that do not cut the poor off from crucial resources—could reduce resource waste and pollution."

Innovator and Catalyst

In approaching the challenge of sustainability, it is useful to keep in mind Einstein's observation that "the world we have created today, as a result of our thinking thus far, has problems which cannot be solved by thinking the way we thought when we created them." Efforts to advance sustainability—in government, the private sector, and civil society—will demand fundamentally new ways of thinking and doing.

Government has a critical role to play in sponsoring research, design, and development in environmentally advanced technologies, and in evaluating and certifying their performance. More broadly, government can play a strategic role in building a strong *capacity for innovation* that supports sustainability. As the WBCSD has pointed out, "Innovation can enable our global economy to depend more on the progress of technology than on the exploitation of nature."

Regulator

When the environment first came onto the radar screen of governments, the focus of public policy was on regulation. Environment policy meant pollution regulation. As the kinds of environmental problems became more complex—for example, less clearly identifiable with single-point sources—and as the policy paradigm broadened from an end-of-pipe focus on environmental pollution to an integrated perspective that encouraged "moving upstream" and encouraged technology innovation, new policy instruments came into the picture. Some began to argue that regulation (to its critics "the old system of command and control") was no longer relevant.

A more balanced view sees regulation as a necessary backdrop for newer approaches to function properly. The challenge for government is to find the appropriate mix of policy instruments and to determine how to make regulation a useful part of that mix. For example, regulation is still needed to ensure minimum performance from environmental laggards, and in some instances to set performance standards in sectors which are critical to the public good.

Organizer of Voluntary and Nonregulatory Initiatives (VNRIs)

Voluntary and nonregulatory initiatives (VNRIs) became popular in the post-Rio period. They cover a range of approaches including compacts, self-imposed targets, and industry codes. They have been heralded by many businesses as effective ways of encouraging action that goes "beyond compliance."

While VNRIs have potential disadvantages such as difficulty in applying the rules to free riders and uncertain accountability, they also have important advantages: flexibility, speed in changing rules based on lessons of experience, low compliance and administrative costs, and avoidance of jurisdictional concerns. They internalize responsibility and make positive use of peer pressure. With transparency and reporting, and civil society organizations serving as verifiers and validators of business voluntary commitments, VNRIs can be highly effective.

Some of the best work on identifying the conditions that must obtain in order for voluntary instruments to succeed has been carried out by a Canadian multistakeholder body called the New Directions Group, which published a report suggesting criteria and principles for effective use of VNRIs.

Box 13.1

Criteria for the Use of VNRIs to Achieve Environmental Policy Objectives

VNRIs should be positioned within a supportive policy and regulatory framework:

1. Interested and affected parties should agree that a VNRI is an

appropriate, credible, and effective method of achieving the desired environmental protection objective.

2. There should be a reasonable expectation of sufficient participation in the VNRI over the long term to ensure its success in meeting its environmental protection objectives.
3. All participants in the design and implementation of the VNRI must have clearly defined roles and responsibilities.
4. Mechanisms should exist to provide all those involved in the development, implementation, and monitoring of a VNRI with the capacity to fulfill their respective roles and responsibilities.

Principles for the Design of VNRIs

Credible and effective VNRIs:

1. Are developed and implemented in a participatory manner that enables the interested and affected parties to contribute equitably.
2. Are transparent in their design and operation.
3. Are performance-based with specified goals, measurable objectives, and milestones.
4. Clearly specify the rewards for good performance and the consequences of not meeting performance objectives.
5. Encourage flexibility and innovation in meeting specified goals and objectives.
6. Have prescribed monitoring and reporting requirements, including timetables.
7. Include mechanisms for verifying the performance of all participants.
8. Encourage continual improvement of both participants and the programs themselves.

New Directions Group, "Criteria and Principles for the Use of Voluntary and Non-Regulatory Initiatives to Achieve Environmental Objectives," 1997. Available from the International Institute for Sustainable Development (IISD) online at http://www.iisd.org/sd/principle.asp.

Educator, Persuader, Information Provider for Decision Making

Governments have the opportunity—and the obligation—to provide information for decision making to all sectors of society. This is an exceptionally broad category of action, with a great deal of room for innovation.

The current OECD *Environmental Outlook* provides an example of the kind of innovation that is urgently needed. It rates recent and projected environmental issues facing OECD members in terms of whether policies developed to date to deal with these issues appear to be adequate (green light), uncertain (yellow light), or inadequate and urgent (red light). Industrial point source pollution, for example, gets a green light, while municipal waste generation gets a red light. This effort provides information in a way designed to encourage governments to explore policy options that could help to alleviate the situation.

Protector of Security

The U.S. Environmental Proctection Agency's National Advisory Council for Environmental Policy and Technology (NACEPT) acknowledges in its recent report *The Environmental Future* that the war on terrorism has "turned the nation's focus to homeland security and the work of disabling international terrorist organizations." Does this mean that the sustainability challenge can now be shifted to the back burner? On the contrary, according to the NACEPT Council:

> Sustainable development is also essential for reducing social unrest and the danger of international terrorism. No mixture of conditions would be more combustible than rapidly expanding numbers of restless young people living in poverty, without opportunities for improvement, constantly exposed to media

images of affluent lifestyles, and influenced by new ideologies that preach hatred against America.

Conclusion

Understood properly, sustainability is not simply a topic to be added to the agenda of governments. It is the lens through which to view the entire agenda—including economic development, social policy, environmental protection, and security—in order to develop integrated, coherent strategies.

Virtually all of the major roles of government can be marshaled into a comprehensive effort to advance corporate sustainability. Pursuing this strategy would limit the role of government by fully recognizing the large roles that the private sector and nongovernment organizations need to play. But it would revitalize government by giving it a central visioning and goal-setting role and integrating its activities on every level around the challenge of facilitating the transition toward a sustainable society that works for everyone, for the long run.

For Further Exploration

David V. J. Bell, "The Role of Government in Advancing Corporate Sustainability." Background Paper prepared for the G8 Environmental Futures Forum in Vancouver, March 2002.

York Centre for Applied Sustainability web site: www.yorku.ca/ycas/.

14

Government in the Chrysalis Economy

John Elkington

From 1960 to the present, three great waves of public pressure have shaped the environmental agenda. The roles and responsibilities of governments and the public sector have mutated in response to each of these three waves—and will continue to do so. Although each wave of activism has been followed by a down wave of falling public concern, each wave has significantly expanded the agendas of politics and business:

- Wave 1 brought an understanding that environmental impacts and natural resource demands have to be limited, resulting in an initial outpouring of environmental legislation.
- Wave 2 brought a wider realization that new kinds of production technologies and new kinds of products are needed, culminating in the insight that development processes have to become sustainable—and a sense that business would often have to take the lead.
- Wave 3 focuses on the growing recognition that sustainable development will require profound changes in the governance of corporations and in the whole process of globalization, putting a renewed focus on government.

The environmental protection role that governments assumed after Wave 1 turns out to be inadequate for supporting the larger economic metamorphosis that now needs to occur. Indeed, the whole concept of "environmental protection" may be limiting our thinking in terms of the necessary scale of change needed for sustainable development. Policies and regulations designed to force companies to comply with minimum environmental standards are inadequate for encouraging the creative, socially responsible entrepreneurship needed to evolve new and more sustainable forms of wealth creation—in what we call the Chrysalis Economy.

Three Pressure Waves

To understand how the roles and responsibilities of government will need to change, we need to consider how the corporations and value chains whose activities governments regulate are themselves evolving through different stages in response to the three waves of public pressure.

The first ("Limits") pressure wave built from the early 1960s. The wave intensified at the end of the decade, peaking from 1969 to 1973. Through the mid-1970s, a wave of environmental legislation swept across the Organisation for Economic Co-operation and Development (OECD) region and industry went into compliance mode. The first down wave followed, running from the mid-1970s to 1987. Acid rain had a major impact on European Union politics in the early 1980s, but this was on the whole a period of conservative politics, with energetic attempts to roll back environmental legislation.

1987 marked a major turning point, however. The second ("Green") pressure wave began in 1988 with the publication of *Our Common Future* by the Brundtland Commission, injecting the term *sustainable development* into the political mainstream. Issues like ozone depletion and rainforest destruction helped fuel a new movement: green consumerism. The peak of the second wave ran from 1988 to 1991. The second down wave followed in 1991. The 1992 United Nations Earth Summit in Rio delayed the impending down wave, triggering spikes in media coverage of issues like climate change and biodiversity, but against a falling trend in public concern. The trends were not all down, however: further spikes were driven by controversies around companies like Shell, Monsanto, and Nike, and by public concerns—at least in Europe—about mad cow disease and genetically modified foods.

The third ("Governance") pressure wave began in 1999. Protests against the World Trade Organization (WTO), World Bank, International Monetary Fund, G8, World Economic Forum and other institutions called attention to the critical role of public and international institutions in promoting—or hindering—sustainable development. The 2002 United Nations World Summit on Sustainable Development brought the issue of governance for sustainable development firmly onto the global agenda—although not onto the agenda of the government of the United States. The United States, which helped trigger and lead the first two waves, has remained in a down wave, running counter to public opinion and pressure in other OECD countries.

Further afield, we expect fourth and fifth waves, very likely on shorter time-frequencies and—possibly—with less dramatic fluctuations in public interest. As these subsequent waves and down waves develop, what we call the Chrysalis Economy will emerge and evolve.

The Chrysalis Economy

A Chrysalis Economy will emerge through an era of intense economic metamorphosis. A key driver will be the unsustainability of current patterns of wealth creation. Today's economy is highly destructive of natural and social capital, and characterized by large and growing gaps between rich and poor. The events of September 11, 2001, serve notice on the rich world that both absolute and relative poverty will be major issues in the future.

Because current patterns of wealth creation will generate continuously worsening environmental and social problems, pressures will continuously build on both corporations and governments to make a transition to sustainable development. Figure 14.1 distinguishes four main types of companies or value webs along the evolutionary path to a Chrysalis Economy: Locusts, Caterpillars, Butterflies, and Honeybees.

The key to developing environmental policies that facilitate the

	Low impact	High impact
Regenerative (increasing returns)	Butterflies	Honeybees
Degenerative (decreasing returns)	Caterpillars	Locusts

Figure 14.1 The MetaMatrix of Corporate Types © SustainAbility, 2001

transition to sustainability is to understand that the roles of government need to be different in relation to the four different types of corporations. For example, corporate Butterflies and Honeybees need to be treated very differently than corporate Caterpillars and Locusts.

Corporate Locusts

Some corporations operate as destructive Locusts throughout their life cycles, others only display locustlike behaviors occasionally. Corporate Locusts everywhere are destroying social and environmental value and undermining the foundations for future economic growth. Some parts of Africa, Asia, Latin America, and regions once controlled by the old Soviet Union are literally crawling with them.

Among the key characteristics of a corporate Locust are:

- The destruction of natural, human, social, and economic capital
- Collectively, an unsustainable burn-rate, potentially creating regional or even global impacts
- A business model unsustainable over the long run

- Periods of invisibility, when it is hard to discern the impending threat
- A tendency to swarm, overwhelming the carrying capacity of social systems, ecosystems, or economies
- An incapacity to foresee negative system effects, coupled with an unwillingness to heed early warnings and learn from mistakes

When most companies were corporate Locusts, government had to take the offensive. Key tasks were to stamp on the worst offenders—and on locustlike behaviors in business as a whole. In a globalizing world, one key challenge for environmental protection agencies is to extend their regulatory and enforcement reach to problem companies operating outside their formal jurisdiction.

Corporate Caterpillars

Usually, Caterpillars are harder to spot than Locusts because their impacts are lower and more localized. But if you live or work right next door to a corporate Caterpillar, their degenerative impacts may make it hard to see that these corporations actually have a significant potential for metamorphosis. Corporate Caterpillars tend to:

- Generate relatively local impacts, most of the time
- Show single-minded dedication to the business task at hand
- Depend on a high burn-rate although usually of forms of capital that are renewable over time
- Operate on a business model that is unsustainable when projected forward into a more equitable world of 8 to 10 billion people
- Have the potential for transformation into a more sustainable guise, often based on a mutated business model
- Operate in sectors in which pioneering companies are already starting to metamorphose toward more sustainable forms of value creation

Here the challenge for governments is to provide appropriate conditions for old businesses to evolve and new businesses to grow, but at the same time use regulatory and financial incentives to ensure that these businesses develop in line with environmental and sustainable development objectives. Key government roles here include:

- Support for research and development
- Technology demonstration programs
- Public/private partnerships
- Green purchasing
- Elimination of perverse subsidies
- Ecological tax reform

Corporate Butterflies

Corporate Butterflies are easy to spot, even though most are comparatively small. By their very nature, they are often highly conspicuous, and in recent years have been abundantly covered in the media (think Ben & Jerry's, Body Shop, Patagonia). An economic system fit for corporate Butterflies would almost certainly be a world well down the track toward sustainability.

Yet even if every company in the world were to model itself on such companies our economies would still not be sustainable. For that, we will need to develop and call upon the swarm and hive strengths of the corporate Honeybee. Even so, corporate Butterflies have a crucial role to play in evolving Chrysalis Capitalism. Among other things, they model new forms of sustainable wealth creation for the Honeybees to mimic and—most significantly—scale up. Some characteristics include:

- A sustainable business model, although this may become less sustainable as success drives growth, expansion, and increasing reliance on financial markets and large corporate partners
- A strong commitment to the corporate social responsibility and sustainable development (SD) agendas
- Often defines position by reference to Locusts and Caterpillars
- A wide network, although not among Locusts or Honeybees
- Increasingly, involvement in symbiotic relationships
- Persistent indirect links to degenerative activities

- A potential capacity to trigger quite disproportionate changes in consumer priorities and, as a result, in the wider economic system
- High visibility and a disproportionately powerful voice for such economic lightweights

Like their natural counterparts, corporate Butterflies tend to occur in pulses. After rain, for example, a desert can suddenly come alive with butterflies. In much the same way, pulses of corporate Butterflies were a feature of the 1960s, with booms in alternative publishing, whole foods, and renewable energy technology businesses, and again in the 1990s, when sectors like eco-tourism, organic food, SD consulting, and socially responsible investing began to go mainstream. Government policies designed to help sound corporate Caterpillars will generally also serve corporate Butterflies well. Government can also encourage change by noticing and celebrating any companies that move from the caterpillar stage to the butterfly stage.

Corporate Honeybees

This is the domain into which growing numbers of government agencies, innovators, entrepreneurs, and investors will head in the coming decades. A sustainable global economy would hum with the activities of corporate Honeybees and the economic versions of beehives. Although bees may periodically swarm like locusts, their impact is not only sustainable but also strongly regenerative. The key characteristics of the corporate Honeybee include:

- A sustainable business model, albeit based on constant innovation
- A clear—and appropriate—set of ethics-based business principles
- Strategic, sustainable management of natural resources
- A capacity for sustained heavy lifting
- Sociability and the evolution of powerful, symbiotic partnerships
- The sustainable production of natural, human, social, institutional, and cultural capital
- A capacity to moderate the impacts of corporate Caterpillars in its

supply chain, to learn from the mistakes of corporate Locusts, and, in certain circumstances, to boost the efforts of corporate Butterflies

Some Implications for Governments

The selective pressures working in favor of sustainable development can only increase. As this occurs, we will see many patterns of change in corporate behaviors. Some companies that remain strongly degenerative will attempt to improve their images through clever mimicry of Butterfly and Honeybee traits. It will not be uncommon to find the same corporation displaying some mix of Caterpillar, Locust, Butterfly, and Honeybee behaviors simultaneously. But no company is fated to remain trapped forever in Locust form. With the right stimulus and leadership, any organization can start the transformative journey, although it is usually easier to go from Caterpillar to Butterfly than from Locust to Honeybee, and from low positive to high positive impact than from high negative to low negative impact.

The roles of government here will be many and various. Aspects of traditional environmental protection approaches will still be needed, but to build truly sustainable wealth creation clusters the public sector will need to take a leaf out of the private sector's book and embark on major "silo-busting" campaigns.

Like corporations and value webs, governments and their agencies will need to move through the various stages shown in the Learning Flywheel (Figure 14.2).

- The first stage focuses on *Invasion*, the natural process by which an innovation—be it a new technology or a new business model—invades an opportunity space, creating economic, social, or environmental impacts in the process. Here government agencies play a key role in identifying new types of impact and pioneering assessment methods.
- In the second stage, we see the emphasis shift to the process of *Internalization*, by which a company or value web absorbs some of the costs previously externalized to society or the environment. Govern-

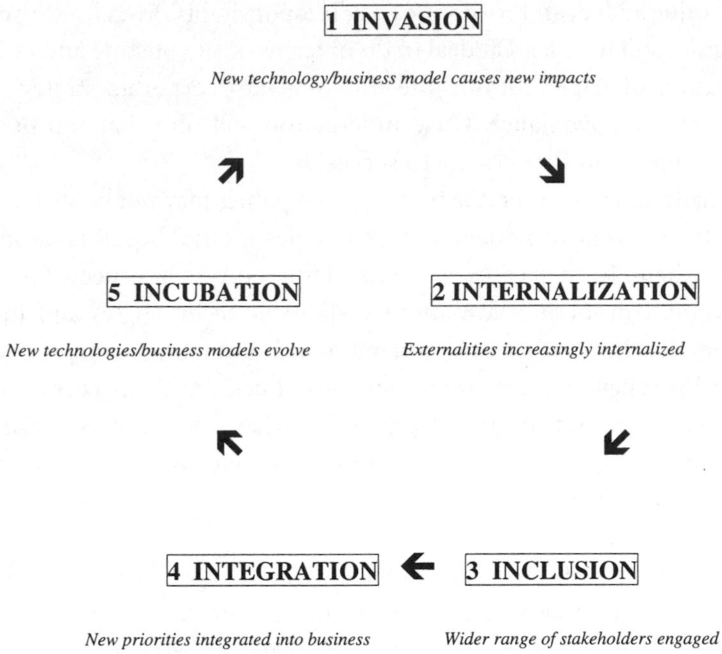

Figure 14.2 The Learning Flywheel © SustainAbility, 2001

ment involvement is critical to ensure externalities are properly costed and internalized.

- As the burdens of internalization build, so management needs to know where the real priorities lie—and we see a new interest in *Inclusion*. This is the process by which a wide range of internal and external stakeholders are progressively engaged, their priorities established, and their legitimate needs met. The public sector has often lagged in this area, but its role will be increasingly significant in establishing key priorities for action and investment.
- Next comes the emerging challenge of *Integration*. Every time business is required to address a new agenda, there is the problem of silos—as has successively been the case with environment, health and safety, total quality management, information technology, sharehold-

er value added, and corporate social responsibility. Even leading companies still have a great deal to do in terms of silo busting and of integration of triple bottom-line thinking into corporate strategy and corporate governance. Governments too will find that silo busting and integration are critical to success.

- Finally, however, even the best-run companies may not be sustainable if their business models or technologies are not sustainable in the long haul. In such cases, we need to focus on the prospects for *Incubation*, considering how more sustainable technologies and industries can be incubated in today's world. Even the most productive beehives have to start from a few brood-cells. And, apart from early projects around industrial ecology, we have hardly even begun to think how governments can catalyze new clusters (geographical or virtual) of sustainable businesses.

In sum, a much more comprehensive approach will be needed that involves a wide range of stakeholders and coordinates across many areas of government policy, including tax policy, technology policy, economic development policy, labor policy, security policy, corporate reporting policy, and so on. Developing this comprehensive approach to sustainable development and environmental protection is the central challenge facing governance in the twenty-first century.

For Further Exploration

John Elkington, *The Chrysalis Economy: How Citizen CEOs and Corporations Can Fuse Values and Value Creation*. New York: John Wiley and Sons, 2001.

SustainAbility web site: www.sustainability.com/home.asp.

15

Is Free Trade Too Costly?

Denis Hayes

In the early nineteenth century, China developed an enormous trade gap with Europe and America. The rulers of the Middle Kingdom were happy to sell silk for silver, but they saw no reason why Chinese should want to buy goods from barbarians. So they banned imports—creating a *real* balance of payments problem.

British traders retaliated by bribing Chinese officials to allow them to smuggle opium into China. The strategy was simple. After the Chinese got hooked on the narcotic, the British could charge more for it without fear of losing repeat sales. The same marketing strategy was recently followed by the Medellín cocaine cartel.

Opium sales quickly caused the British trade gap to disappear. Seeing this, most American trading companies immediately entered the opium business. Opium sales by Warren Delano, an American businessman, formed much of the basis of Franklin Delano Roosevelt's family fortune. Warren Delano wrote home to his wife that he could not pretend to justify the opium trade morally, "but *as a merchant* I insist it has been . . . fair, honorable, and legitimate."

By 1839, opium addiction had become epidemic in China. Finally, the Emperor began to enforce the law. Vast stocks of opium were seized in

Canton and dumped into the sea. In response, Britain waged and won the First Opium War.

You may be thinking, "This is all sort of interesting, but what's the point?"

The point is that this kind of behavior now seems morally repugnant to us. Americans take some pride today that we have left behind the amoral—no, let's be candid, the immoral—extremes of nineteenth century capitalism. The excesses that Charles Dickens described still can be found today in scores of early-stage capitalist societies in places such as Indonesia and Peru and Russia. But, here at home, Americans have tempered most of the excesses of cowboy capitalism. We have banned child labor and established universal public education. We have legislated a forty-hour work week and a minimum wage; outlawed trusts and monopolies; standardized financial reporting; safeguarded occupational health and safety; and prohibited racial, sexual, and religious discrimination in the workplace.

In recent decades, a new social value—a safe, healthy environment—has emerged as a major concern for Americans. Moreover, Americans have decided that protecting the environment is an arena in which state intervention is needed. Because environmental protection often carries a price tag, no company can afford to behave responsibly unless it is assured that its competitors will be forced to meet the same environmental standards. Enlightened business leaders have often been strong supporters of tough laws that would apply to all. The role of government—domestically and internationally—is to ensure that irresponsible companies do not get an unfair advantage over responsible firms by dumping their externalities into the air or streams.

Although the American economy bears no resemblance to a sustainable economy, we have made striking progress over the last three decades, particularly in the realms of air and water pollution. However, many international investments remain strikingly unchanged from the nineteenth century.

World trade is still dominated by a mind-set that the man on the street finds appalling. A major attraction of overseas investment often is the opportunity to pay wages that barely approach subsistence levels, and to emit toxics that in this country would lead to prison terms. In a famous memo, Larry Summers—then chief economist of the World Bank—argued that it is economically most efficient for rich countries to dump their toxic wastes in poor countries because poor people have shorter life spans and less earning potential than rich people.

Where Does the Environmental Movement Fit into This?

The world, as never before, is interconnected. Even as New England seeks to control sulfur emissions from upwind power plants in the Midwest, nations have a legitimate interest in the environmental behavior of other sovereign states. Greenhouse gases emitted in Russia affect the climate in Africa. Ozone-depleting chemicals used to clean computer chips in Europe contribute to skin cancer rates in Florida. As a result, no country can go it alone in environmental protection, and environmentalists have to be concerned with the operations of the global economy, including the dynamics of international trade.

Many prominent environmental groups have developed deep reservations about the way the current international trade regime operates. The reasons for their concern have not been well explained to the public. Like most Americans, environmentalists do not want an international trade regime in which it is legitimate to addict populations to opium or dump toxic wastes in poor countries. They believe that the balancing act between maximum production and social values has been a central component of American greatness and needs to be extended to the global economy.

Free trade zealots dismiss environmental objections by painting environmentalists as pastoral isolationists. But using pastoralist arguments to portray the environmental concerns about trade is like using libertarians who favor free trade in narcotics as emblematic of the trade community. Such caricatures are mostly efforts to avoid addressing real issues.

As an example of a real issue, consider paragraph 4 of Article 16 of the document that created the World Trade Organization (WTO) during the Uruguay Round of General Agreement on Tariffs and Trade. It states, "Each member shall ensure the conformity of its laws, regulations and administrative procedures with its obligations as provided in the annexed Agreements."

Any law requiring imported goods to meet local or national health, safety, labor, or environmental standards that exceeds WTO international standards can be overturned by the WTO—even though the country applies the same or tougher standards to its own domestic manufacturers. Standards regarding carcinogenic food additives, automobile safety, toxic substances, meat inspection, pesticides, or even informational labels are all subject to challenge when they exceed WTO standards. And WTO standards, almost without exception—indeed, almost by definition—are set at the lowest common denominator.

Similarly, measures that restrict the export of a country's own resources—such as logs, minerals, or fish—can be ruled unfair trade practices. Requirements that locally harvested timber be processed locally before export in order to provide local employment almost certainly could not survive WTO review.

When Taiwan proposed a law that would ban cigarette sales in vending machines, restrict public smoking areas, prohibit all forms of tobacco advertising, and fund a public education campaign to encourage people to give up smoking, the U.S. Trade Representative threatened to call for trade sanctions against Taiwan—even though those laws would have affected Taiwanese tobacco companies and U.S. imports equally. This contemporary action involving a dangerous, highly addictive export is difficult to distinguish morally from Warren Delano's comments about opium.

Or consider the case in which a U.S. company raising vegetables in Mexico uses a pesticide that leaves a toxic residue that complies with international standards but does not meet the standard of the state of California. If the company persuades the Mexican government to bring an action against the California standard under WTO, California would have no right to appeal an unfavorable WTO ruling in any court.

The WTO panel possessing this extraordinary power consists of three trade experts, typically corporate lawyers, who meet in secret. Its recommendations are automatically adopted sixty days after presentation to the WTO unless there is a *unanimous* vote of WTO members to reject them. It is hard to conjure up a process more inimical to the aspirations of international civil society. Three appointed corporate lawyers, meeting in secret, can invalidate laws passed by Congress and signed by the president. In fact, they have done it, in a case involving dolphin-safe tuna. *More than one hundred countries, including the country that brought and won the decision, must vote unanimously in order to overturn the panel.*

Of course, the United States gets to have input into setting the WTO standards. But this, too, offers little reassurance to environmentalists. A study in 1991 found that of the 111 members of the U.S. Trade Representative's three main trade advisory committees, only 2 represented labor unions; one approved seat for an environmental advocate remained unfilled; and there were no consumer representatives. Ninety-two members represented individual companies, and sixteen represented industry associations. Among the ninety-two companies represented, twenty-seven had paid fines totaling more than $12 million to the Environmental Protection Agency (EPA) between 1980 and 1990, and five made the EPA's top ten list of hazardous waste dischargers. The panels also included scores of major funders of antienvironmental lobbying campaigns, and backers of antienvironmental political candidates. The advisory panels rarely announce their meetings to the public, and never allow the public to attend.

Can anyone be seriously surprised that environmentalists, who have fought hard for thirty years to mobilize public support to clean up our air and water, reduce toxics, and protect endangered species, are appalled at the transfer of de facto power to these new forums that we are almost powerless to influence?

The Big Question

The core environmental concern with free trade, however, is the vision of the future promoted by WTO policies: endless material growth with no acknowledgment of the need for a transformation in the nature of growth. Few oppose trade in principle, and few oppose growth in principle. But both trade and growth need to be guided by a vision more *visionary* than simply a quest for more.

During the first half-century of George W. Bush's life, Americans consumed more of the world's mineral wealth than all people in all societies throughout the entire course of history before his birth. Everyone is familiar with the bromide that Americans, with 5 percent of the world's population, consume about 30 percent of the world's resources. We ignore the corollary: if everyone consumed at the American level, our oil reserves would shrink to just a few years' supply, global warming would cease to be an abstraction, and all the world's old-growth rainforests would disappear swiftly. If 1.2 billion Chinese consumed like Americans, the environmental impact would be multiplied to five times as great as America's. The natural life-support systems of the earth would quickly collapse.

Twentieth-century America is not sustainable in its current form, and it cannot be broadly replicated. If the world is to enjoy prosperity, we need *a different model of what prosperity means*, and an international trade regime that promotes this new model.

What is needed is illustrated by progress in the realm of computing. ENIAC, the first serious electronic computer, used to dim the lights in one-third of Philadelphia whenever it was turned on. Today, we can pack vastly more computational power into something the size of a cornflake and run it for hours on a small battery.

Just as it has been possible to do more and more computation with less and less electric power, fivefold, and even tenfold, improvements in the energy and materials intensity of lights, appliances, buildings, vehicles, and industrial processes can become commonplace.

One can envision an attractive world in which the recycling of basic metals approaches 100 percent; in which paper is routinely recycled several times before being consumed as fuel; in which all energy is derived from renewable sources powered directly or indirectly by the sun; in which healthy, low-meat diets are within the biological carrying capacity of the planet; in which information-dense, super-efficient, pollution-free technologies guide commerce, transportation, and residential living; in which the greatest new surge of economic growth is targeted in environmentally friendly hardware and software associated with telecommunication, computation, entertainment, and the other pillars of the information revolution.

What America Can Offer the World

If the world has an attractive future, it will be one in which a stable human population makes superefficient use of materials and energy. No model currently exists of such superefficient, sustainable prosperity. Simple economic growth certainly will not take us there. Unbridled global markets currently are taking the whole world rapidly in the opposite direction. Trade and development can be—must be—hugely positive forces. But that statement describes an aspiration, not a current reality.

America can act on that aspiration. We can help create a new model of sustainable prosperity, and our economy can thrive by using international trade to spread this model around the world. Indeed, design efficiencies, coupled with decentralized energy sources and the decentralized networking allowed by cellular technology and satellite technology, will soon allow some daring, developing country to leapfrog the industrial age. Instead of spending decades and most of the national treasury developing a massive, unneeded infrastructure of roads, railroads, ports, transmission lines, and so forth, this country will jump directly from a mostly agricultural economy to the information age. (Important elements of this leapfrogging can already be found in places like Singapore, Bangalore, and South Korea.) U.S. aid programs, innovative international banks, and enlightened companies ought to be alert for candidates, and be prepared to generously help them prove this alternative development model.

Finally, when we talk about international trade today, we are generally talking about the exchange of physical goods. But the most important benefits of international exchange are probably in the realm of ideas. While guarding against cultural imperialism, we should not be bashful about marketing certain ideals at least as aggressively as we market computers and airplanes. The American dream stands for something more than raw commerce.

What I have in mind are the fundamental guarantees of our Bill of Rights: free expression, religious tolerance, democratic self-determination, cultural freedom, the rule of law, uncompromising opposition to political persecution and torture, and policies and actions to preserve and enhance the ecological integrity of the planet.

If we don't hold ourselves to high standards in these areas, and if we don't intelligently, surgically use our influence to promote them abroad, then we have shortchanged the importance of the great, ongoing experiment that Thomas Jefferson helped launch two hundred years ago.

> We hold these truths to be self-evident, that all men [*all people everywhere*] are created equal, that they are endowed by their Creator with certain unalienable Rights, that among these are Life, Liberty and the pursuit of Happiness. . . . That to secure these rights, Governments are instituted among Men, deriving their just powers from the consent of the governed. . . .

That—not twelve-cylinder recreation vehicles to drive to the shop-till-you-drop malls, *that*, not *Survivor VII*, Britney Spears, and fancy running shoes—*that* is what America is all about. That is the most important thing we must nourish and protect at home. That is the most important gift that we can offer the world.

For Further Exploration

Bullitt Foundation web site: www.bullitt.org/.

Denis Hayes, *The Official Earth Day Guide to Planet Repair*. Washington, D.C.: Island Press, 2000.

16

Greening the Global Financial System

Hazel Henderson

From species extinction and climate change to spreading deserts and conflicts over water, the environmental challenges we face today are increasingly global. The most critical economic challenges before us are also increasingly global. The growth produced by a more open world economy is failing to "trickle down," and poverty gaps within and between countries are widening precipitously. The rapid growth of global markets and trade has taken down the firewalls between countries' economies and created a new global casino—a 24/7 seamless financial and currency-trading system. Without global accounting rules and prudential regulation of stock markets, the growing volume and velocity of these hot money flows can lead to massive instabilities, loss of domestic macroeconomic management options, bouncing currencies, and turbulent financial markets. Environmental challenges and the challenges of creating a more equitable and stable global economy come together in the concept of sustainable development. And there is a growing body of evidence and opinion that a better approach to globalization, including fundamental reform of the global financial system, is needed to foster global sustainable development.

Worldwide protests are beginning to emerge calling on politicians and business elites to take up the task of global reform. The common media term "antiglobalization protests" is a gross mischaracterization. Very few people really oppose the globalization of human interactions, which is impossible to stop in any case. The great majority of people involved in these activities are actually grassroots globalists who believe that today's technological globalization must be accompanied by global institutional reform, enlightened national policy, and a new level of global corporate social responsibility to insure that globalization occurs in a way that reduces poverty, protects the environment, and promotes democracy.

The ideological underpinning of the present Washington Consensus approach to globalization is the paradigm of zero-sum, competitive, neoclassical economics. The self-maximizing behavior, deregulation, and privatization favored by this approach work well in allocating resources in open systems with few human agents operating in resource-rich ecosystems. The problem is that as we humans have multiplied to our current population of 6 billion and colonized most of the planet's ecological niches, the earth's formerly wide open spaces have transformed gradually from free goods—air, watersheds, oceans, and biodiversity—of economic theory into commons that can only be used and managed with rules, agreements, and private or public property regimes. What we are seeing today is the early stage of the emergence of new non-zero-sum, win-win strategies for globalization and cooperative sustainable development.

Today, U.S. capital markets are suffering a crisis of investor confidence—precisely because the underlying winner-take-all, competitive economics paradigm has collided with the new reality. Today's $1.5 trillion daily currency markets and related stock, bond, commodity, and futures markets are a *global commons* resting on trust, truth (i.e., accurate accounting to investors), and confidence in the ethics and fairness of the players and the rules of the game.

Corporate leaders, investment bankers, security analysts, accounting firms, ratings agencies, asset managers, brokers, dealers, stock exchanges, and the politicians they support with campaign donations urgently need to understand the deeper implications of economic globalization: *greater transparency, honesty, cooperation, and new rules for the commons are necessary as a framework for the successful operation of global markets and competition.*

This understanding is at least beginning. For example, former secretary of the Treasury Lawrence Summers came to support a number of win-win rules for the global playing field such as tighter G8 and Organisation for Economic Co-operation and Development (OECD) regulation of offshore banking, tax havens, and money laundering. Small economies and Switzerland, which competed to offer phantom incorporations, numbered bank accounts, and freedom from regulations, taxes, disclosure and oversight, were ostracized. Prior to September 11, 2001, President Bush pulled the United States out of this G8-OECD program, which had led to many voluntary reforms. Soon after September 11, Bush and his Treasury Secretary, Paul O'Neill, learned the lessons of the new global economic commons and rejoined this cooperative agreement—as the best way to track and curtail the funds al Qaeda was funneling to terrorist groups.

Similarly, associations of bankers, stock exchanges, and trade associations have begun to voluntarily promulgate global standards. Even competitors in telecom and other high-tech sectors are learning the lessons of cooperative standards setting. They discovered their chaotic competition of differing standards and protocols for all phones and other equipment had allowed the Europeans to race ahead with cooperative standards U.S. companies' products could not meet.

Much deeper changes are needed, however, to foster sustainable development and spread the benefits of globalization to reach the 2 billion members of the human family still living in deprivation, ill health, ignorance, and despair. One of the critical requirements for a future that works is fundamental reform of the global financial system to promote the globalization of human rights, social justice, environmental sustainability, and opportunities for human development.

The Earth Charter: A Framework for Global Reform

The deeper changes we need have been expressed clearly and eloquently in the sixteen principles of the new United Nations (UN) Earth Charter (www.earthcharter.org), which is widely viewed by international lawyers as the equally important companion treaty to the Universal Declaration of Human Rights. The principles of the Earth Charter provide a comprehensive framework for the emerging global debates about global standard setting, reshaping international economic institutions, and reforming the world's financial architecture.

Global standard setting is advancing in the areas of human rights, workplace standards, and environmental protection and restoration—whether voluntary or via binding protocols and treaties. So far, this is necessarily a fragmented process because it involves values embedded in many cultures, institutions, academic curricula, business practices, and voluntary, nonprofit civic society organizations and movements. The Earth Charter provides a broader framework based on over a decade of multicultural, multilateral consultations, which can serve as a benchmark and starting point for many of today's standard-setting activities.

Current efforts to overhaul national accounts to include human, social, and environmental capital and quality of life can also benefit from the reference points of the Earth Charter. The traditional scorecards of "wealth" and "progress"—gross national product and its narrower version, gross domestic product—are no longer adequate, and broader scorecards are becoming available. Preeminent is the Human Development Index (HDI) pioneered by the United Nations Development Programme and released annually since 1990. The HDI covers poverty gaps, environmental quality, health, relative budget priorities between military spending and education, and other aspects of government performance in over 180 countries. The Calvert-Henderson Quality of Life Indicators provide a systems view of sustainability trends in the United States.

Hot Money and Ecological Destruction

Short-term hot money flows (currencies and portfolio investments) have become transmission belts of ecological destruction and the exacerbation of poverty. These financial flows are far more crucial to the sustainable development agenda than trade—since they dwarf the 10 percent of global trade-related transactions in the $1.5 trillion daily totals of currency exchange. Ninety percent of these daily flows are speculative and unrelated to trade.

Proposals for currency exchange taxes seek to address this problem. At the UN Social Summit in Geneva, in June 2000, 160 governments agreed to perform feasibility studies on currency exchange taxes—although the United States refused to participate. Currency exchange taxes and other policy innovations were driven off the agenda of the UN Summit on Financing for Development held in Monterrey, Mexico, in March 2002—primarily, again, by the United States. But these ideas will not go away. Estimates of revenues from even a 0.01 percent currency exchange tax range from $50 billion to $300 billion annually. Great opportunities lie ahead for using such innovative funding sources for global health promotion, environmental protection and restoration, and other aspects of cooperative global development.

Greening the Financial System

G8 finance ministers and central bankers have repeatedly called for a "new global financial architecture" since the Asian financial crises of 1997. Even after crises that followed in 1998 involving Russia and then Brazil, Turkey, and Argentina, official rhetoric has not been matched by action and results. After severe public criticism by such mainstream economists as Jeffrey Sachs and Joseph Stiglitz, the International Monetary Fund (IMF) has taken some responsibility for its macroeconomic policy prescriptions. These orthodox policies and their conditionalities exacerbated the 1997 crises in Thailand, Indonesia, Korea, and other Asian countries. They required draconian cuts in social programs, further opening of these economies and the like, causing runs on their currencies and plunging millions into poverty.

The performance of the Bretton Woods twins (the IMF and World Bank) has been no better regarding reducing or canceling the unrepayable debts of heavily indebted poor countries. Many of these debts are deemed "odious," that is, they were incurred in corrupt deals between politicians and their corporate and financial cronies—and should be repudiated. The rapid reduction of unrepayable debt is necessary—

but not sufficient to build a basis for alternative, sustainable paths to development. It may be necessary for many indebted developing countries to seek bankruptcy protection. The most appropriate model is not Chapter 11, as now proposed by Anne Kreuger of the IMF, but Chapter 9 of the U.S. Bankruptcy Law, which covers municipal bankruptcies. Chapter 9 allows the continuation of all social programs, services, and public expenditures and is therefore a way to protect the vulnerable and poor of a country seeking protection under this law.

Both the IMF and the World Bank need to be redirected, democratized, and restructured for more limited missions and made transparent and accountable to all countries—not only their rich shareholders. The IMF should desist from its dangerous fixations regarding opening up economies and their capital accounts before their financial sectors, public institutions, industries, and civil societies are robust enough for global competition. The World Bank should relinquish involvement in structural adjustment and in the financing of large infrastructure and centralized energy projects. The World Bank's focus should change to direct funding of grassroots initiatives such as AIDS prevention, education, distributed energy generation, micro-enterprise development, and other projects that clearly improve the lives of ordinary people rather than enriching economic elites.

Beyond international regulation of global finance, markets, and trade, new institutions are needed. These include a *World Environment Organization* to balance the narrow focus of the WTO and an *International Bank for Environmental Settlements* (as proposed by the United Nations Development Programme in Paper 10 of 1997) to manage the disputes and inequities arising from global climate change and to organize the contraction and convergence approach to equitable per capita emission rights in a trading system with deep liquidity for economic efficiency.

Global Governance of Transnational Corporations

Of the one hundred largest economic entities in the world, half are now transnational corporations (TNCs) and the other half are nation-states. That means that the largest fifty TNCs surpass in size the economies of

130 of the 180 recognized states of the United Nations. TNCs increasingly rival or surpass nation-states in terms of global impact. Irresponsible behavior by such powerful entities can hinder global development. But TNCs also have a global reach and outlook and an unparalleled capacity for innovation and technology transfer that could be redeployed toward fostering sustainable development.

Just as government regulation of national corporations proved necessary in national economies, so now better global regulation of TNCs is essential for pursuing the global agenda of promoting human rights, democracy, sustainable development, and poverty eradication. Current approaches include international treaties on human rights, labor standards, and environmental protection. A more multipronged effort to contain and redirect the productive and destructive powers of TNCs will need to include:

- Changing the charters of TNCs from their legal requirement to maximize shareholders' returns on their investments to reflect the interests of all stakeholders: employees, suppliers, consumers, host-communities, society at large, governments, and the environment. Many TNCs are chartered in nations, principalities, or provincial-level states (as in the United States) that compete to offer TNCs the most lax charters with the fewest requirements for auditing, transparency, and even protection of shareholders. A global Multilateral Agreement on Investors and Corporate Responsibility for best-practices chartering and protocols, rooted in international law, is needed. States could also offer such best-practices charters to TNCs wishing to reposition themselves in sustainability sectors.
- Expanding international accounting standards along the lines of the Global Reporting Initiative, which promotes triple bottom-line accounting and corporate annual reports (i.e., economic, environmental, and social accounting). Progress is being made in this area by the movements of socially responsible investors, which in the United States alone hold $2.1 trillion of the shares of companies that pass such triple bottom-line accounting standards.
- Building support for the United Nations Global Compact launched by Kofi Annan in 1999 at the World Economic Forum in Davos. The

Global Compact invites TNCs to engage with its nine principles of good corporate citizenship in human rights, labor standards, and environmental protection, but is voluntary and lacks compliance or performance criteria. However, the signatory companies are receiving additional scrutiny by nongovernmental organizations and importantly by socially responsible asset management firms.

- Financing an expansion of public-access, noncommercial media at all levels from global to local as a counterbalance to TNC media dominance. Today we live in "mediocracies" where a handful of Northern media barons dominate global television, movies, radio, newspapers, book publishing, video games, and increasingly the Internet. These media serve the global TNC marketplace that buys advertising, creating a strong yet often unconscious bias against views that challenge existing global economic arrangements.

National governments have a vital part to play in cooperatively creating the global governance mechanisms needed to circumscribe TNCs and global financial markets. Most of the international taxes and innovative funding discussed by civic groups at the UN Summit on Financing for Development would be collected by national governments. These revenues can offset capital flight and tax evasion, bolster nations' currency-stabilization funds, and fund selective international agencies for humanitarian purposes or development. Simply eliminating perverse subsidies to unsustainable activities and economic sectors, estimated at some $800 billion to $1 trillion annually worldwide, would save more than the $650 billion annually estimated in *Agenda 21* as the cost to shift societies toward sustainability. National tax codes must be shifted to taxation of waste, pollution, planned obsolescence, and virgin resource extraction—to jumpstart recycling, reuse, remanufacturing, and barter sectors. Revenue neutrality requires a concomitant reduction in taxes on incomes and payrolls.

Finally, global peacekeeping also falls within the context of reforming international economic institutions and is arguably the most urgent requirement for sustainable human development. Here again, the Earth Charter provides a point of departure. The world of twenty-first century interconnectedness dictates win-win, cooperative approaches for mutually managing conflict in the many global commons we humans unavoidably share.

The proposal for a United Nations Security Insurance Agency (UNSIA) is an example of the kind of win-win innovation needed in this area. The UNSIA would be a public-private, civic partnership between a reformed UN Security Council, the insurance industry, and the hundreds of civic humanitarian organizations in conflict resolution and peace building. Any nation wanting to cut its military budget could apply to UNSIA for a peacekeeping "insurance policy." The insurance industry would supply the political risk assessors and write the policies. The "premiums" would be pooled to fund properly trained peacekeepers and rapid deployment of the existing networks of civic and humanitarian groups. The UNSIA proposal is backed by several Nobel Peace Prize winners and is already being seriously discussed in many universities.

Protecting the environment and improving the process of globalization are more closely related tasks than most people—including environmentalists—have realized. Fundamental reform of the global financial system will ultimately be necessary to foster global sustainable development. The Earth Charter's sixteen principles provide basic benchmarks to address the central issues of global reform.

For Further Exploration

Hazel Henderson, Jon Lickerman, and Patrice Flynn (Eds.), *Calvert-Henderson Quality of Life Indicators*. Bethesda, MD: The Calvert Group, 2000. Updates online at www.calvert-henderson.com.

H. Henderson, *Beyond Globalization: Shaping a Sustainable Global Economy*. Bloomfield, CT: Kumarian Press, 1999.

H. Henderson, *Building a Win-Win World: Life beyond Global Economic Warfare*, New York: Berrett-Koehler, 1996.

17

The Challenge Ahead

Bob Olson and David Rejeski

Many years ago, someone questioned the management guru Peter Drucker about his uncanny ability to predict the future. Drucker answered that "I never predict the future, I just look out the window and see what is visible, but not yet seen." This collection of essays is a look out the window. It presents nothing in the realm of science fiction but provides a glimpse of a number of puzzle pieces yet to be assembled. In this last section, we will attempt to put some of these pieces together, both in summary and through the use of scenarios.

The first section of this book examined a sampling of major technological developments underway today, including advances in resource productivity and energy, genomics, nanotechnology, pervasive computing, and global manufacturing. There are many other equally important areas of technological change, from the evolution of the Internet and new techniques of restoration ecology to advances in fuel cells and hydrogen production and storage. These developments are all likely to have major environmental impacts, for good, for bad, or for both. And many of these developments are interacting, with progress in one area stimulating developments in others. The result is a technological acceleration even more far reaching than the acceleration of change that occurred in the industrial revolution of the nineteenth century, with profound implications for the environment.

Today, if we look out the window, we can begin to discern the outlines of the next industrial revolution—a once-in-a-century chance to shape an emerging technological and organizational system so it works with the environment, not against it. The exact outlines of this revolution are hidden behind a complex screen of jargon like mass customization, value chain modularity, contract manufacturing, distributed manufacturing, build-to-order, real-time enterprise, personalization of production, free-agent workers, transgenics, and molecular assembly. Behind this gibberish, however, is a fundamental change in the way we produce, where we produce, and whether we even chose to produce, or increasingly substitute information for matter. When the fundamental nature of production changes, so must environmental protection.

The first industrial revolution produced massive environmental impacts. The environmental and social consequences of today's technological acceleration are likely to be even larger. The technological developments unfolding today harbor great promise, but also great dangers. They could unleash enormously destructive impacts on the environment and society, but they can also be the basis for an environmentally sustainable civilization. As was the case with the last industrial revolution, these impacts (both positive and negative) will not be distributed equitably across our planet. Large segments of the global population will fail to benefit from environmental advance unless governments act on their behalf.

During the nineteenth century, observers as diverse as George Perkins Marsh, John Muir, Henry David Thoreau, and Harriet Lawrence Hemenway commented on how new technologies were having a disruptive impact on the natural world. But no organized efforts foresaw these impacts in advance, and no legal and administrative infrastructure, no tools of governance, headed these impacts off. Today we have many more tools, but the question is their adequacy in the face of new sets of emerging challenges. The book's second section dealt with the potential for improvements in governance. Technology may be a powerful driver of change, but the direction of change is not predetermined, and governance and our collective ingenuity will be the ultimate deciders.

The challenge ahead is to harness emerging technological capabilities to the vision of creating a humane and sustainable future. That requires making deliberate social and technological choices using all the tools of governance, defined broadly to include not just the formal institutions of government, but also learning and decision processes within private corporations, nongovernmental organizations (NGOs), and the networks that interlink institutions of all kinds. In democratic societies, governance is ultimately based on the perceptions, values, and actions of large numbers of citizens.

Meeting this challenge will require going beyond environmental protection as we have known it. The environmental movement and the major public institutions responsible for environmental protection have spent much of the last thirty years passing legislation and enforcing regulations designed to deal with the damages of a century-old revolution in industrial production. This has been a successful effort on the whole. But now the emphasis needs to shift from reducing the environmental impacts of an old technological order to creating more advanced technological systems that eliminate most environmental impacts altogether. Environmental protection needs to go beyond reactive responses and become part of a more proactive effort to shape the technological infrastructure of a sustainable future. This will be an especially difficult transition, since we have, in effect, become experts at the old ways of dealing with problems, and old habits and routines die hard.

Three Scenarios of Technology and Governance

The following three scenarios illustrate the different ways technology and governance could coevolve to shape the environmental future over the coming decades:

- In the first scenario, Old Ways, outmoded approaches to governance give little support to new technologies important for achieving sustainability, and sometimes actively block their emergence. Older, environmentally damaging technologies remain dominant long into the future and transitions to new technological regimes are started too late or badly managed. Older technologies and old approaches to governance remain effectively locked in.

- In the second scenario, Catch-Up, new technologies emerge at an accelerated pace, some of which produce positive environmental gains, but governments fail to foresee and meet head-on the potentially serious environmental problems many new technologies pose. Governance, though improved or reinvented, is always too little too late, with environmental protection efforts only coming into play after new technologies are already widely deployed and producing massive impacts.
- In the third scenario, New World, a technologically advanced and more sustainable world is shaped by new approaches to governance. Better foresight and cooperation between sectors make it possible to guide technological progress in order to produce positive environmental outcomes and minimize unintended consequences.

Scenario 1: Old Ways

From the standpoint of public policy, sometimes the easiest way forward is to go backwards. This scenario requires no new money and little political risk taking, and offers a comfortable organizational home for tired bureaucrats. It involves doing "less with less"—less vision, less innovation and entrepreneurship, and less leadership. If this scenario has a ring of the familiar to it, it should. We have been living it in the United States for the last four years.

Box 17.1

Old Ways

Outmoded Governance Slows Environmental Improvements

- Little or no environmental leadership or vision in any sector
- Influenced by old industries, government tilts the playing field even more in their favor

- Continued high reliance on traditional regulation, and many cases of regulatory rollback
- Adversarial relationships between government, business, and NGOs
- Low international cooperation and rapidly worsening global environmental problems
- Reductionist analytical strategies applied to emerging systems problems
- Old approaches to environmental protection discourage technological innovation and "lock in" polluting technologies

Backsliding to less demanding old ways is easy, but if we continue on this path the consequences will be hard on the environment and will ultimately undermine economic development. If new technologies and strategies critical for reducing environmental impacts fail to emerge, the result will be rapid global climate change, a massive global loss of biodiversity, and acute water scarcities that undermine both agriculture and urbanization. If renewable energy technologies and more efficient technologies for using energy are not deployed in time, global warming will not be the only problem. Economic development in poorer nations will be torpedoed as escalating global oil consumption exceeds global production capacity and rich nations outbid the poor for access to increasingly expensive fossil fuels and other key resources. It will be a future of shrinking forests, collapsing fisheries, declining biodiversity, and spreading deserts. The invaluable ecosystem services we have always taken for granted—the cycling of water and nutrients, natural controls on diseases and pests, the assimilation and detoxification of wastes—will begin to fail. The real environmental progress that has been made in the United States and many other nations will be overwhelmed by the larger disruption and breakdown of global ecological systems.

Avoiding a future like this requires discontinuous, coordinated changes in investment and policy to catalyze what Gus Speth calls "an environmental revolution in technology." As Lester Brown puts it, "we need to reach agreement as rapidly as possible on the need for systemic change. . . ." But government usually changes in a disjointed and incremental manner, and tends to resist systemic change or simply muddle through. Large shifts in direction require the kind of visionary leadership described by Joanne Ciulla, where leaders are willing to educate the public and challenge society with bold goals. That kind of leadership sometimes emerges but has always been rare, and it is certainly in short supply today.

Moreover, many desirable technological changes will be actively resisted. Mature industries built around previous generations of technology have the political muscle to influence government research and development spending, subsidies, tax policies, and regulations to keep the playing field tilted in their favor. Over the past generation, for example, more than 90 percent of all government energy subsidies, totaling in the hundreds of billions of dollars, have gone to fossil fuels and nuclear power—not an approach likely to foster an environmental revolution in energy efficiency and new sources of energy.

Environmentalists themselves may sometimes block the development of environmentally advanced technologies, if they are not careful in their assessments. For example, a number of NGO groups are increasingly focusing on the potential dangers posed by biotechnology and nanotechnology. This is important to do, but also important to do well. Exaggerated alarms about nanotechnology, for instance, which fail to discriminate between its different forms, could discourage the development of the kind of applications Mark Wiesner and Vicki Colvin describe, such as using nanomaterials for detecting contaminants, turning toxic wastes into harmless byproducts, improving fuel cells, and increasing the efficiency of photovoltaic cells.

Even environmental protection, as a field, contributes little to encouraging an environmental revolution in technology. Environmental protection typically focuses on technologies to address specific environmental problems, like catalytic converters for reducing auto emissions. It gives little attention to understanding the much larger environmental opportunities inherent more generally in technological change—opportunities in many areas to design out potential problems and develop environmentally positive applications. Technology-based regulations often tend to freeze in existing technologies and discourage companies from pursuing technological innovations.

Old Ways is therefore an all too plausible scenario. We are living in it right now, and many of the dynamics in government, industry, and even the environmental community will tend to keep society on this path.

Scenario 2: Catch-Up

Suppose innovation and entrepreneurship break through the barriers that limit change in the Old Ways scenario. Developments in information technology, biotechnology, nanotechnology, and other areas converge in a new industrial revolution. Suppose regulatory rollback is stopped and commitments to public sector reinvention take hold. Some positive results would certainly follow, but our society would be confronted with another major challenge—the challenge of keeping up with rapid technological change. Generally, our ability to introduce new products and processes expands exponentially, while our ability to understand their impacts grows arithmetically, at best.

Box 17.2

Catch-Up

Governance Fails to Keep up with Rapid Technological Change

- Speed of government action is slowed by political conflicts and outmoded organizational designs
- New sets of technologically induced environmental problems appear and surprise governments and the public
- Inadequate foresight mechanisms (dominant "short-termism")
- High reliance on traditional command and control mechanisms
- NGOs and corporations drive environmental progress more than government
- Low level of international engagement by the public sector, degradation of global environmental quality

- Weak public sector leadership (frustrates leadership efforts in other sectors)
- Few systemic approaches to problem understanding and problem solving

We live in a cleaner country, free of belching smokestacks and burning rivers, since the advent of the modern environmental movement around 1970. Nevertheless, our society has always played catch-up with the environmental problems caused by emerging technologies. A future where society falls further and further behind is entirely possible, because novel, technologically induced environmental problems are likely to appear with growing frequency, and the hardest problems of the past will not go away (think of the number of Superfund sites that still remain to be cleaned up).

A growing mismatch exists between the speed of government, slowed by drawn-out battles among parties and between Congress and the administration, and the acceleration of technological change, propelled forward by global economic competition, innovation, and scientific advance. What Brad Allenby calls the "new foundational technologies" such as nanotechnology, genetic technologies, and information technologies pose the greatest challenge. To get a feel for the mismatch one only needs to compare various science budgets. The federal research budget for the life sciences has doubled since 1998. In the area of nanotechnology, the new U.S. government funding initiative will pump close to $1 billion annually into this area for the next four years. On the other hand, the U.S. government investment in environmental science has been essentially flat for over twenty years.

The potential problems being explored by Mark Wiesner and Vicki Colvin, like the persistence of nanoscale particles within cells, may be just the leading edge of a host of novel risks associated with emerging technologies. Gary Marchant predicts that new genetic technologies will have enormous environmental impacts, for good or for ill, radically transforming the landscape of environmental issues and policy. Timothy Sturgeon suggests that "traditional approaches to environmental protection tied to vertical organization" will be "consigned to history" by the information systems restructuring global manufacturing.

Rapidly evolving and powerful technologies like these threaten to produce an ongoing series of unpleasant environmental surprises. Because the political process is so slow, regulatory standard-setting and other efforts to address the problems are unlikely to come into play until the technologies causing the problems are widespread and the impacts on the environment and human health are already high, and sometimes irreversible.

While new technologies are generating novel environmental problems, an agenda of unsolved environmental problems keeps piling up. We have picked most of the low-hanging fruit off the environmental improvement tree. What remains are more diffuse and distributed problems, involving myriad small sources with large aggregate impacts. And, as Gus Speth points out, the most difficult environmental problems of all are the global ones that can only be solved through cooperative global action.

Really catching up with environmental problems requires foresight to identify potential problems early and head them off before they become crises. But society's foresight mechanisms are inadequate, and, as Stewart Brand warns, "civilization is revving itself into a pathologically short attention span." Catching up with environmental problems also requires government to take on the range of roles described by David Bell, including steering change by articulating visions and setting long-term goals. It requires investing heavily in the approach advocated by William McDonough and Michael Braungart that aims to eliminate toxic emissions altogether—by cradle-to-cradle design.

By statute, structure, and bureaucratic inertia, however, environmental protection remains unable to take on such roles in an effective way. Weak public sector leadership and a low level of international engagement confine environmental protection to a sharply limited sphere of influence.

Given these realities of governance, a Catch-Up scenario must also be considered quite plausible. The future could well be some mix of the Old Ways and Catch-Up scenarios.

Scenario 3: New World

The first two scenarios may be highly plausible, but neither is desirable or inevitable. A far better future—a New World scenario—will only be possible if we envision it clearly. Considerable progress has already been made in this direction. The idea of a *sustainable society*—a society with enough foresight, flexibility, and wisdom to avoid undermining the ecological and social foundations on which it is built—has already emerged as an influential "image of the future." The idea of an *environmental revolution in technology*, in which advancing technology is guided by humane and environmental values, is replacing the old polarity between uncritical technological optimists and technocritics skeptical of all advanced technologies. Environmental policy is beginning to move beyond business-as-usual toward more anticipatory approaches to environmental protection and more comprehensive strategies for supporting the development of environmentally advanced technologies.

Box 17.3

New World

Transformational Technologies Combine with Transformed Governance

- Hybrid governance strategies using the right mix of regulation, markets, and networks
- Strong, functioning international regime ensuring progress on global issues and the integration of environmental goals into international development assistance
- High level of organizational learning and knowledge transfer between sectors enabled by new technologies and organizational strategies
- Institutionalized foresight mechanisms (the "long view") to develop long-term goals and accountability measures
- Strong corporate environmental responsibility and sustainability movement

- Technological progress shaped to produce positive environmental outcomes and minimize unintended consequences
- Especially dangerous lines of technology development are identified early, discussed thoroughly, and avoided
- Strong environmental leadership in all sectors and institutionalized leadership training
- Advance to systems thinking and management strategies that look beyond a single medium

Continuing to move in this direction will require improvements in governance—improvements beneficial for society as a whole as well as for the environment. Improving governance is a difficult but manageable challenge for the generation ahead. Most of the needed improvements are a matter of mindset, culture, and intention. However, new organizational structures will be needed and, invariably, some new expenditures. The use of "skunkworks" to provide the freedom and space for new ideas and creative thinkers will be key to this scenario.

Here are some of the immediate challenges and some no-lose strategies for improving our ability to steer the next industrial revolution toward sustainability. First, the idea of sustainable development needs to be elevated to a higher level than environmental protection. As John Elkington argues, environmental protection is too narrow a concept. Its focus on policies and regulations to force companies to comply with minimum environmental standards is inadequate for encouraging the creative, socially responsible entrepreneurship needed to bring about an environmental revolution in technology. It is not a concept that can, by itself, inspire new approaches to development that will be successful over the long run economically and socially while dramatically reducing environmental impacts.

Second, far more effort in government needs to be devoted to improving foresight to identify potential environmental impacts of emerging technologies and to interacting with business in the early stages of technology development to help design out negative impacts and support environmentally positive applications. A growing number of efforts of this kind have been introduced within the U.S. Environmental Protection Agency (EPA) in areas like nanotechnology, genomics, fuel cells, radiation protection, and indoor environments. However none of the efforts undertaken to date have had a significant impact on agencywide priorities or strategic plans. Eventually, as the EPA's Science Advisory Board said in its 1995 report *Beyond the Horizon*, "As much attention should be given to avoiding future environmental problems as to controlling current ones."

Third, the human resources devoted to understanding emerging developments in technology need to be upgraded. The pervasiveness, speed, and complexity of emerging science and associated technologies are exceeding the capacity of the environmental community to respond. Organizations are already being simultaneously pulled in multiple directions by disruptive changes in biology and computer science. Given the enormous public and private sector investments in nanotechnology, we can expect extremely rapid innovation and unanticipated spillover effects, which will add to, and interact with, effects from the info and biotech realms. Especially hard hit will be the NGOs, who are otherwise occupied fighting unending battles to stop regulatory rollbacks and other stealth maneuvers by the barons of the last industrial revolution. Many local, state, and federal environmental organizations will not fare much better, as they will have to compete with the private sector for people with the skill sets to operate in these new areas or in the interstitial spaces between them (such as in bio-computation).

In his 1986 science fiction novel *Count Zero*, William Gibson lays out a future world in which the battles are not between nation-states fighting for land, money, or resources, but between organizations vying for talent and creativity on a global scale. The public sector needs to enter that battleground or become irrelevant. The environmental workforce in government has aged over the past thirty years and needs to be evaluated and restructured to make sure that agencies have the human, not just financial, resources to deal effectively with new challenges both in and across these emerging and converging disciplines.

Fourth, the front line of environmental protection needs to shift from the legal department to the science and technology functions. If we are at a critical juncture in our industrial evolution, then there is only one viable strategy in this situation, to proactively shape the future, a function that our existing regulatory infrastructure is not well suited for.

This does not portend the end of environmental law and regulation. However, part of the legal profession must position itself at the front of the technological curve. It is urgent that we carefully examine the existing regulatory framework in terms of adequacy to deal with emerging science and technology. This will require a deep, not superficial, analysis across the regulatory landscape within agencies, across agencies, and across geographic boundaries (local, state, federal, and international). The task will be made more difficult because innovation will be occurring between, rather than in, the disciplines and sectors where traditional laws and regulations have been developed and tested. Regulatory gaps will need to be identified and filled and the transparency of the regulatory system constantly improved, especially for small businesses that may be driving innovation in emerging sectors.

Fifth, agencies such as the EPA and its equivalents around the globe will need to retool their research strategies. Too much funding is still being spent dealing with the last industrial revolution and its aftermath and by-products, and not enough on preparatory and anticipatory research. Given the level of scientific and technological innovation taking place at this point in time, funding at the EPA for so-called exploratory research is unacceptably low and not directed in a strategic enough fashion. Funding should include robust programs focused on societal and ethical implications in areas such as toxicogenomics.

The need is urgent to develop potential breakthrough technologies with research and development funding targeted directly at producing positive disruptive change (not a 3 percent improvement in efficiency or reduction in cost, but change by a factor of three or more). This is the way the Defense Advanced Research Projects Agency (DARPA) has traditionally functioned within the Department of Defense. That's the agency that gave us the Internet. Now is the time to create a DARPA-style office within the EPA (and EPA equivalents worldwide) to tackle the really hard problems with unorthodox approaches. How much money should such an office receive? Between 1995 and 2003, DARPA's funding averaged 5.3 percent of total Department of Defense research and development. A 5 percent figure applied to the EPA's existing research and development budget would result in over $30 million devoted to the search for game-changing technologies. The driving ethos of such a project should be, as Apple computer founder Steve Jobs once said, to "put a dent in the universe." Such an office or department should become a magnet for the most creative talent in the world.

Finally, in an era of pervasive scientific change, we need pervasive scientific literacy, and that includes our public, our press, and our policymakers. We can expect the complexity of the science underpinning both environmental problems and solutions to continue to increase, demanding ever more sophisticated understanding transcending multiple disciplines. Over a decade of survey research done by the Roper Center Public Opinion Research for the National Environmental Education and Training Foundation showed that as complexity of environmental issues increases, public understanding drops off precipitously. A scientifically illiterate public will be extremely susceptible to various scare campaigns in the press, films, or other media. Nanotechnology has become the poster child for technohype as it creeps into the public consciousness through advertisements, TV shows, books, and films. In this environment, it will be harder for the public to separate science from science fiction and harder for the government and corporations to engage in meaningful dialogues with public constituencies over future issues.

Our ability to prepare society for the next industrial revolution is closely related to our ability to perceive and anticipate change and understand its implications for present actions and policies. Far too few resources in the environmental community are dedicated to understanding the changing context in which policies and strategies will be developed and implemented. Some future historian may well characterize this point in our environmental history as one of great tragedy, not only because of the unenlightened attacks on our existing environmental laws, but also because we missed an opportunity to reshape our industrial infrastructure in ways that would make it far more environmentally benign and sustainable. In a recent interview, Sun Microsystems' former chief scientist Bill Joy noted that "we need to encourage the future we want, rather than try to prevent the future we fear." Too many times, environmental protection has been focused on fears rather than aspirations. We need to break that habit, and the opportunity is now.

The Challenge Ahead

These scenarios are not predictions of the future; they are tools for thinking more clearly and creatively in situations in which the future cannot be predicted. They provide an intellectual framework and a common vocabulary for strategic conversations about what could happen, what we want to happen, and how to make things happen.

Clearly the Old Ways and Catch-Up scenarios are highly plausible. In fact, recent government activity to roll back or delay regulations (in the case of mercury, for instance) indicates that these scenarios are in force. But the New World scenario is possible and by far the most desirable. Since governance will play such a large role in determining what happens, we need strategic conversations within the institutions of governance to explore in more depth what the New World scenario entails and what actions could be most effective in moving toward that future.

A future in which we fail to shape emerging technologies for social and environmental gain is a future in which unintended consequences and spill-over effects are both more likely and potentially more pernicious. If we fail to make the necessary investments in science, technology, and human capital to achieve the New World scenario, then we will need to significantly expand institutional mechanisms to mitigate the unwanted effects of technological advance. The costs of dealing with major new environmental problems would be far greater than the costs of heading problems off at an early stage.

The New World scenario cannot be achieved though bipartisan bickering, weak leadership, or unilateral strategies. The environmental challenges before us will not bend to half-hearted, incrementalist strategies, let alone to the complete lack of long-term strategy dominating today's environmental dialogue. But that is only half the story.

The other half is just as compelling: incrementalism seldom motivates human imagination or attracts talent. In today's world, highly creative people have better places to invest their idealism, energy, and know-how than in tired public sector institutions pursuing mediocre objectives. Because technology is only a means to an end, extracting the most from our technological future depends on extracting the most from human imagination. That is unlikely to happen until we are charged with some bold, audacious environmental goals.

As George Bernard Shaw observed, "progress depends on the unreasonable man [or woman]." If environmental progress has slowed, maybe it is because we have become too reasonable in our expectations and too comfortable with the old technological order to understand and shape the new. Maybe the only way to meet the challenge ahead is by reaching for a New World.

About the Authors

Brad Allenby is the vice president for Environment, Health and Safety at AT&T; an adjunct professor at Columbia University's School of International and Public Affairs and the Princeton Theological Seminary; and a Batten Fellow at the University of Virginia's Darden Business School. He is the author of numerous books and articles including *Industrial Ecology: Policy Framework and Implementation* and, recently, "Observations on the Philosophic Implications of Earth Systems Engineering and Management." He received his law degree from the University of Virginia and a Ph.D. in environmental sciences from Rutgers University.

David V. J. Bell is professor and former dean in the Faculty of Environmental Studies at York University and director of the York Center for Applied Sustainability. He is the editor of the Sustainable Development theme of the Encyclopedia of Life Support Systems published by the United Nations Educational, Scientific and Cultural Organization. He has worked with a number of departments and agencies in the Government of Canada and recently prepared background papers for the G8 Environmental Futures Forum in Vancouver and for the Earth Summit 2002 Canadian Secretariat.

Stewart Brand is a social inventor as well as author of *The Clock of the Long Now*, *How Buildings Learn*, *The Media Lab*, and *The Whole Earth Catalog*. He is a founding member of the Long Now Foundation, and he cofounded the Global Business Network. Both organizations are dedicated to fostering "the long view."

Michael Braungart is a German chemist who is the cofounder and principal of McDonough Braungart Design Chemistry, LLC, a product and systems development firm assisting client companies to implement sustainable design protocols.

John Seely Brown is the former chief scientist of Xerox, and former director of its Palo Alto Research Center. He has been responsible for guiding one of the most famous technology think tanks in the world, and leading one of the most celebrated and far-ranging corporate research efforts. Part scientist, part artist, and part philosopher, Brown's viewpoints are unique and distinguished by a broad view of the human contexts in which technologies operate, as well as a refreshing skepticism about whether or not change always represents genuine progress.

Lester Brown is founder and president of the Earth Policy Institute, an organization dedicated to promoting a vision of an eco-economy. He is one of the world's most influential thinkers in the area of environmentally sustainable development, a concept he helped to pioneer. During a career that started with tomato farming, Brown has been awarded over twenty honorary degrees and has authored or coauthored forty-eight books, nineteen monographs, and countless articles. He is a MacArthur Fellow and the recipient of many prizes and awards, including the 1987 United Nations Environment Prize, the 1989 World Wide Fund for Nature Gold Medal, and the 1994 Blue Planet Prize for his "exceptional contributions to solving global environmental problems." He is perhaps best known for having founded the Worldwatch Institute in 1974 where he served as president for its first 26 years, launching the Worldwatch Papers, the two annual reports *State of the World* and *Vital Signs*, and *World Watch* magazine.

Joanne B. Ciulla is Professor and Coston Family Chair in Leadership and Ethics at the Jepson School of Leadership Studies at the University of Richmond. She is one of the founding faculty of the Jepson school. Ciulla has a B.A., an M.A., and a Ph.D. in philosophy. Her books include *The Ethics of Leadership*, *The Working Life: The Promise and Betrayal of Modern Work*, and *Ethics, the Heart of Leadership*. She is currently coauthoring a book called *Honest Work*.

Vicki Colvin is an associate professor in the Department of Chemistry at Rice University and Director of the Center for Biological and Environmental Nanotechnology. Prior to her association with Rice, she was a member of the technical staff at Bell Labs where she developed new materials for holographic data storage. She received her Ph.D. in 1994 at the University of California—Berkeley. Her undergraduate degrees, B.A.'s in chemistry and physics, were completed in 1988 at Stanford University. She is the author of over twenty-five refereed publications, three patents, and one book chapter.

John Elkington is cofounder and chair of SustainAbility (www.sustainability.com), based in London and New York. He is a leading authority on sustainable development and triple bottom-line business strategy. His latest book is *The Chrysalis Economy: How Citizen CEOs and Corporations Can Fuse Values and Value Creation.*

Denis Hayes is president of the Bullitt Foundation, a $100 million environmental philanthropy located in Seattle. He also serves as chairman of Earth Day Network, which is now active in 180 nations. During the Carter administration, he was director of the federal Solar Energy Research Institute. He has also been a professor of engineering at Stanford University, an attorney with the Cooley Godward firm in Silicon Valley, a Visiting Scholar at the Woodrow Wilson Center, and a senior fellow at the Worldwatch Institute. Hayes was national coordinator of the first Earth Day in 1970. Time magazine chose him as a Hero of the Planet, and the Sierra Club gave him its John Muir Award.

Hazel Henderson is an independent futurist, economist, and consultant on sustainable development. She is the author of *Beyond Globalization* and seven other books. Her editorials appear in twenty-seven languages and more than four hundred newspapers. She serves on several boards including the Calvert Social Investment Fund, the Worldwatch Institute, and the Cousteau Society.

Amory Lovins is a physicist and an energy expert. He is the founder and CEO of Rocky Mountain Institute. His work focuses on transforming the electricity, automobile, real estate, water, semiconductor, and other industries toward advanced resource productivity. His concept of a "soft energy path" restructured the energy debate in the 1970s. He has received a MacArthur Fellowship, the Heinz, Lindberg, Right of Livlihood ("Alternative Nobel"), World Technology, and *Time* magazine Hero for the Planet awards as well as nine honorary doctorates.

Hunter Lovins is currently director of the Natural Capitalism Group, a consultant on sustainability for business and the nonprofit sector. She is a cofounder of the Rocky Mountain Institute and served for many years as its CEO for strategy. A lawyer and member of the California Bar, she helped establish and for six years was assistant director of the California Conservation Project, an innovative urban forestry and environmental education group. She has coauthored nine books, including *Natural Capitalism*.

Gary Marchant is an associate professor at Arizona State University College of Law and the executive director of the Azizona State University Center for the Study of Law, Science and Technology. He received a Ph.D. in Genetics from the University of British Columbia in 1986, a Masters of Public Policy from the Kennedy School of Government at Harvard University in 1990, and a J.D. in 1990 from Harvard Law School. His teaching and research interests include environmental law; risk assessment and risk management; genetics and the law; and law, science, and technology.

William McDonough is an internationally renowned architect and designer. He is the founding principal of William McDonough+Partners

Architecture and Community Design, a design firm practicing ecologically, socially, and economically intelligent architecture and planning in the United States and abroad. He recently served as dean of the School of Architecture at the University of Virginia. He is the co-author of Cradle To Cradle, a design blueprint for what he and his partners call "The Next Industrial Revolution."

Robert Olson is Senior Fellow and a founding board member of the Institute for Alternative Futures. He has been a project director and consultant to the director at the Office of Technology Assessment of the U.S. Congress, a fellow at the American University's Center for Cooperative Global Development, and an adjunct professor of Public Policy at George Mason University. He served as co-chair of the Transportation and Infrastructure working group of the President's Council on Sustainable Development.

David Rejeski is director of the Foresight and Governance Project at the Woodrow Wilson International Center for Scholars and was recently a visiting fellow at Yale's School of Environmental Studies and Forestry. He headed the Futures Studies Unit at the Environmental Protection Agency, served at the White House Office of Science and Technology Policy, and directed the Environmental Technology Task Force at the White House Council on Environmental Quality.

David Ronfeldt is a senior political scientist working in the International Security and Policy Group at RAND and a leading expert on networks. He is the author (with John Arquilla) of *In Athena's Camp: Preparing for Conflict in the Information Age* and *Networks and Netwars: The Future of Terror, Crime, and Militancy.*

Gus Speth is currently dean of the Yale School of Forestry and Environmental Studies and recently headed the United Nations Development Programme. He founded the World Resources Institute and served as its president for eleven years. Earlier he cofounded the Natural Resources

Defense Council and served as its senior attorney. He chaired the White House Council on Environmental Quality during the Carter Administration.

Tim Sturgeon is a research associate and executive director of the Industrial Performance Center's Globalization Study at the Massachusetts Institute of Technology (MIT). Prior to this, Sturgeon served as Globalization Research Director for the International Motor Vehicle Program at MIT's Center for Technology, Policy and Industrial Development.

Mark Wiesner is the director of the Environmental and Energy Systems Institute at Rice University and currently serves as director of the Shell Center for Sustainability. He is a professor in the Departments of Civil and Environmental Engineering and Chemical Engineering at Rice University. Dr. Wiesner's research pioneered the application of membrane processes to environmental separations and water treatment and recently initiated an examination of the fate and transport of nanomaterials in the environment.

Feng Zhao is a Senior Researcher and directs the Networked Embedded Computing Group at Microsoft Research in Redmond, Washington. He is also a Consulting Associate Professor of Computer Science at Stanford University, and serves as the founding Editor-in-Chief of the ACM Transactions on Sensor Networks. Previously, he was a Principal Scientist and managed the sensor network research at Xerox PARC. He has been investigating how to program and organize large collections of interconnected devices such as wireless sensors. He received a Ph.D. in Electrical Engineering and Computer Science from MIT in 1992. He was an Alfred P. Sloan Research Fellow, and received a National Science Foundation Young Investigator Award and Office of Naval Research Young Investigator Award.

Index

Island Press Board of Directors